Urban Landscapes

Massimo Sargolini

Urban Landscapes

Environmental Networks and Quality of Life

Springer

Massimo Sargolini is Associate Professor in Town and Regional Planning at the School of Architecture and Design "E. Vittoria", University of Camerino

ISBN 978-88-470-2879-1 ISBN 978-88-470-2880-7 (eBook)

DOI 10.1007/978-88-470-2880-7

Springer Milan Dordrecht Heidelberg London New York

Library of Congress Control Number: 2012952990

9 8 7 6 5 4 3 2 1 2013 2014 2015 2016

Cover design: Deblik, Berlin, Germany
Typesetting: Graphostudio, Milan, Italy

Springer-Verlag Italia S.r.l. – Via Decembrio 28 – I-20137 Milan
Springer is a part of Springer Science+Business Media (www.springer.com)

Foreword

New Visions in a Changing World

The great economic and financial crisis triggered at the beginning of this century seems to deny a place for a rational reflection on **contemporary territories**. The imperative for "growth," on which the attention of different social entities converges, crushes any worry for the **environment**, the **landscape**, and quality of life. Efforts to leave the crisis and face the risks that characterize the events under way, bend to the logic of the emergency, turning our attention away from problems and events of everyday life and ordinariness.

Additionally, the intrinsic nature of the **global crisis** constrains us to question the structural roots and long-term evolutionary tendencies, the convergence of earthly upheavals of uncommon violence, the disastrous effects of global climate change [1], dramatic geopolitical aftershocks on the international scale, and the explosion of the "speculative bubble." The awareness of the responsibility that current society assumes toward future generations in terms of "deferred costs," both social and environmental, irreversible processes of degradation, and "planned calamities" (the inevitable fruit of incorrect territorial planning choices), slowly creeps in. It is necessary to bring back into discussion the development models and interpretive paradigms that have guided economic/territorial dynamics over the last century. The international debate is now rich with signals in this direction, as testified by recent International Union for Conservation of Nature (IUCN) declarations, in particular at the World Conservation Congress held in Jeju, Korea in 2012, which witnessed an explosion of clamorous protests toward large international sponsors. These signals allow us to glimpse the irreversible crisis of "Western" political matrices in the environmental movement through the eruption of new immense continental realities and the growing social sharing of basic ethics and equity demands.

If we look at the responses to the crisis that technical and scientific culture has produced in recent decades, we realize a sense of profound inadequacy, which in large part descends from the "sectoralization" and partiality of those responses. The crisis of "modernity" [2] manifest over the course of recent decades had signaled the decadence of the great narratives, unifying ideas of progress and social justice, and general questions, such as the "**urban question**" in all its forms. Fragmentation of the ecosystem produced by total urbanization of the territory seemed to find comparison in the chipping away of the social fabric and relationship systems, in the extreme diversification of interests, stakes, and social demands.

It is now necessary to ask how to move beyond this vision, if possible; if the exit from the crisis pervading contemporary territories requires, once again, the construction of a wide and comprehensive interpretational framework, in which one can try to coherently seat the problems of **contemporary society**, from those that concern the physical transformation of the planet to those that reflect the radicalization of economic and social inequalities, the lack and degradation of living conditions, the growing mobility of populations, as well as activities and ideas; if and how the culture of the territory today—in terms of scientific statutes and disciplinary skills, professional technical capabilities, administrative apparatus, judicial instruments, institutional frameworks, and cultural attitudes and sensitivity—can translate "*the individual involvement in public questions*" [3].

Places and Networks

In the attempts by many experts to respond to these questions, the present volume recognizes the importance of theoretical work and research programs that coalesce into two concepts evocative of the title itself: **places** and **networks**. Thought of, for some time [4], as a complementary metaphor for the interpretation and project of contemporary territoriality, the two concepts have been dealt with frequently under different disciplinary profiles, giving rise to new paradigms that can help us understand the changes under way.

On one side are the paradigms that stress local values and the reasons for **local development**, threatened or attacked by current models of development and more generally by the globalization of economic, social, and cultural dynamics. In the "territorialist" approach—such as proposed by the Italian "Society of Territorialists" [5]—the local territory is placed at the center of disciplinary attention as a common good in its historical, cultural, social, environmental, and productive identity:

> *The return to the territory as a cradle and as the result of human action expresses and symbolizes the need to reintegrate within the social, and therefore economic, analysis, the effects, always determined historically and geographically, of human action on the human mind and the natural environment [5].*

The territory, therefore, is not an inert supporter of anthropic activities, but a system consisting of relationships that are entertained with the natural dynamics.

However, this same concept of **identity** has been the object of growing criticism. In fact, on the one hand, the identity demands have not lacked a show of regressive tendencies, "autistic" closure, and nostalgic drifting behind the bars of tradition. On all scales, from the local to the planetary, the shadow of "armed" or "bellicose" identity has advanced, threatening [6]. On the other hand, the "scaling up" of a growing number of environmental, economic, and social problems (always less treatable at the local scale, always requiring more controlling actions and apparatus at the regional, national, or international scale), brings up again the great themes of the trans-scale integration of public regulatory action, grand accounts and big pictures on which collective strategies can be founded. The changes in the scale of problems to be faced verify the distortions of identity, since they look not only at real changes in the contemporary world, but also at the ways in which we observe them, at the ideas, myths, and utopias that guide our actions and choices.

The revitalization of the **local dimension** has found significant support in the European Landscape Convention [7], which emphasized three aspects: the extension of conservation politics to the entire territory, the complex meaning of the landscape, and the involvement of local actors. The definition of "**landscape**", conceived not only as the dynamic outcome of the interaction among anthropic and natural factors, but also as "foundation of the identity," has found wide application in discussion and project practices in recent decades, in which the landscape has often been the flag of the defense and reinforcement of local cultures and economies, even in resistance to the standardizing push of globalization. As a system of identity values, the landscape not only interprets the uniqueness, character, and differences of the territory, but also its evolutionary potential. From here arises the growing importance attributed to the landscape, with the aim of defending biodiversity and, even more so, "**biocultural diversity**", which the United Nations Educational, Scientific, and Cultural Organization and the IUCN have recently called attention to again. From here also arises the attempt to contribute, with landscape policies, to the rediscovery and improvement of local cultures, agriculture, and economies, placing them at the center of urban and territorial marketing.

It is in the local dimension that landscapes are created, with dynamic and trans-scale processes more or less explicitly linked to the **perceptions**, expectations, and projects of the inhabitants and of those who take care of the landscapes. But places are not autonomous, independent fragments, nor are they separate worlds; they are "*splinters of the world*" [8]. Their capacity to conserve their own identifying characteristics depends on "openness" toward change, more than on the "operational closing" of local systems, and therefore on their capacity to effectively face challenges in the network of relationships acting on the territorial context on different scales.

In both directions, the landscape carries out a fundamental role. On the one hand, it links tangible and intangible facts, places, and activities variously dispersed in the territory to basically cohesive contexts, in which **local communities** recognize themselves and interact more or less effectively. The landscape apparatus is the external manifestation and tool of ecological, economic, social, and cultural *cohesion*, designed to reinforce the resistance and resilience of local systems toward pressures and perturbations coming from the exterior. But at the same time, the landscape is also the external manifestation and tool of *competitiveness*, to the degree in which it confers on local systems a characteristic and recognizable image that allows local entities to participate with the hope for success in confrontations outlined at all levels, mobilizing structured combinations of diversified resources, giving voice to the "territory of the inhabitants," or investing in urban and territorial marketing.

It is here that the landscape paradigm intersects the **reticular paradigm**, or better, that wide range of conceptual multidisciplinary frames that refer to the concept of network: from traditional and innovative concepts of road systems, transportation, and communication; to those that connect business systems and productive, commercial, or energetic activities; to those that recall complex, stratified historical/cultural, religious, or commemorative **relationships**; to those that, over the centuries, have constructed the articulated urban framework with which we still measure; to those that now reflect the emergence of the "**diffuse**" or "**branching**" **city** [9]. These reticular concepts have been called to confront or interact with ecological or landscape ideas ever more frequently, giving rise to biocultural hybridization. Internationally, the construction of ecological networks, aimed at counteracting or reducing ecosystem fragmentation, has often assumed different scopes that are more complex than the original, strictly biological, ones [10]: against the route of bears or wolves are placed the "song routes" of the Australian aborigines or the routes of the transhumance of pastures. In the reticular city, which is profiled in new urban horizons, connection networks cannot hold only biological functions, but must inevitably assume a denser and more complex meaning that integrates nature and culture, connecting different resources and values. One notices ever more the need to create a new multidimensional "environmental infrastructure" capable of stimulating the entire territory, undertaking a supporting role that is no less important than what is traditionally assigned to transportation, communication, or energy networks.

Cities and Urban Landscapes

The realization of effective **biocultural networks** implies ecosystem approaches designed to reconstruct the relationship between nature and culture; a relationship that should keep all scales in mind, from an incompatibility that should be respected on the one hand, to ecosystem interactions that should be promoted or protected on the other ("*separate when necessary, inte-*

grate wherever possible"*). This is especially important for the policies of parks and protected areas, to avoid or reduce insularization and **improve** their economic and social impact ("Benefits beyond boundaries", as was recommended at the Fifth World Parks Congress in Durban [11] in the "new paradigms" framework). However, that double need applies more generally to the management of territories in which we operate, rich with diversity and potential conflicts deriving from centuries-old processes of anthropization: "historical" territories in all respects, where it is necessary above all to "denaturalize" the changes under way, discovering their unavoidable anthropic responsibility, and not concealing them behind a generic warning about events and false natural calamities.

Redesigning the relationships between *naturality*, *rurality*, and *urbanity*, recognizing their pervasive coexistence in any corner of the planet, signals a radical conquering of that counterposition between nature and culture that has held a central role in Western society in the modern era, starting from the great Renaissance utopias, moving to the "pre-urbanistic" proposals of the 1800s (such as the Theory of the Three Magnets by E. Howard [12]), and then reaching the manifestos of the "modern movement" in the first half of the 20th century. The invention of landscape orients the gaze and irreversibly humanizes the natural world, even in the absence of important physical transformations. The naturality that finds space in interpretations and landscape devices is what was historically determined by prior anthropic appropriation of the space: not only that which left us an admirable heritage of cultural landscapes, but also the homogenizing push that has invested the country, hyper-simplified and trivialized the agrarian landscapes, and bent natural dynamics to the logic of settlement dispersion [13]. Such "historicized" naturality is not in any way confinable to "natural areas" detached (in an illusory way) from anthropic influence, but intersects the work of humans everywhere, whether it becomes a secondary ally to natural dynamics, or curbs areas in more or less calamitous ways.

Couched in these terms, the question of quality of life can find meaningful responses in the two paradigms recalled in this text: the *landscape paradigm* and the *reticular paradigm*. The first, to the degree in which the social notion of landscape builds a bridge between nature and culture, between identity and universal values, between factors of cohesion and factors of competitiveness. Together, they can allow the rights of citizenship to be obtained, the goals of conservation to be achieved, and the individual and collective fruition of **heritage** and common goods (tangible and intangible) to be guaranteed, thereby defining the conditions of living in local contexts. The second, to the degree in which the reticular interpretation of territorial dynamics can allow for counteraction of the isolation and closure of local communities, and for the promo-

* This phrase was used in 1985 as the title of a meeting between the Politecnico University of Turin and the University of Delft, and then continued in several papers of the author in the management of river belts.

tion of connections between systems of resources, activities, and fruition that are spread differently throughout the space.

Both paradigms encounter their main area of application in the city. This is not only because quality of life in the city represents the living conditions for most people today, but also because "*in a long-term perspective, cities are designed and equipped like a large functional and territorial infrastructure in support of society and the economy*" [14]. It is in the city that we find the key to recognizing the territoriality of the contemporary landscape, its criticality and its potential [15]. It is in the landscape that the city interacts not only with the "built" countryside [16], but also with the mobile presence of **nature**, which resistant to any confinement, pervades and diversifies the surrounding space. Despite diffuse urbanization processes, the proliferation of the mesh of infrastructure, and the same "modernization" of cultivation practices that have profoundly and extensively eaten away at landscapes inherited from the past, shattering their identifying characteristics, it is still in these landscapes and their consistent evolution that we can try to recover the quality, beauty, and recognizability of contemporary territories.

The importance of **urban landscapes** is destined to be accentuated in the current transition phase toward the society of culture and knowledge, in which the role of the city is always less entrusted with the classic functions of the tertiary and quaternary sectors, and ever more with symbolic relationships, identifying images, and "intangible" dynamics. It is here that landscape "implementation" comes into play, which communicates and emphasizes urban centrality and its significance in contemporary society as a specific level of urbanity, the ultimate essence of that "**right to the city**" on which the urban struggles of the 1970s were founded [17], but also newly at the center of the social demand for the city and memory. The idea is being discussed that only that space for convergence and physical aggregation—the square—can host values of centrality, and that, on the contrary, open spaces, the country, and spaces of "urban dissolution" are dialectically opposed. In new urban configurations, the great events more frequently take shape outside of traditional places; everyday life appropriates unusual spaces; youth continually reinvent spaces for creativity. The rearticulation of **urban centrality** does not look at pieces of the city or fragments that are detachable from the territory, but rather views the historical territory as a whole.

If the city changes its relationship with the territory, it also changes the role of free and open spaces, always less interpretable via the ambiguous metaphor of "urban greenery," which are ever more frequently the theater of new urban phenomenology. Here, free spaces can branch out dynamically within the territorial networks, entering and leaving the compact city (typically with rivers and water systems), crossing and linking it to the territorial context. In this framework, the "**renaturalization**" of the city, the object of recurring urban and territorial politics on the international level, can take on a new and different meaning. It is considered not so much to concede additional space to the urban greenery, as to bring nature back to the city, restoring its full ecological, historical, and cultural meaning. The efforts that are being made in so many European cities to rethink and focus on the idea of "**green belts**," within new perspectives of the reticular city branching out into the territory, are testament to the need to identify a new organizational rationale capable of integrating open and closed spaces, urban and rural landscapes, settlement and environmental dynamics.

In this direction, nature and landscape conservation policies seems to challenge the urban planning culture, forcing it to go beyond its traditional constraints, and to take on new, weightier responsibilities concerning the creation of public spaces and the protection of common goods. But the challenge also more generally looks at the territorialist culture, called to express knowledge, interpretations, and projects that consider both the interests and values in play, going beyond the current frustrating separation between institutions, tools, and forms of public regulation used in the different sectors. In this perspective, the alliance between policies for **nature conservation** (and in particular for the creation of environmental networks) and **landscape policies** represents a crucial necessary passage.

Roberto Gambino
Full Professor in Town and Regional
Planning at the Politecnico di Torino, Italy

References

1. Kushner J A (2008) Global Climate Change and the Road to Extinction. Caroline Academic Press, Durham
2. Harvey D (1990) La crisi della modernità. Il Saggiatore, Milan
3. Bauman Z (2008) La solitudine del cittadino globale. UE Feltrinelli, Milan
4. Gambino R (1994) Luoghi e reti: nuove metafore per il piano. In: Archivio di studi urbani e regionali, 51, Milan
5. Magnaghi A (2010) Manifesto della Società dei territorialisti (in progress)
6. Amartya S (2006) Identità e violenza. Laterza, Bari
7. COE, Council of Europe (2000) European Landscape Convention, Florence
8. Magris C (1997) Microcosmi. Garzanti ed, Cernusco sn
9. Dematteis G (1995) Progetto implicito, Franco Angeli, Milan
10. CED-PPN European Center for Documentation on Natural Parks Planning (2008) Parchi d'Europa. ETS, Pisa
11. The World Conservation Union (2003) 5th World Parks Congress: Benefits beyond Boundaries, Durban
12. Howard E (1898) Garden Cities of Tomorrow, London
13. Gambino R (2000) Reti ecologiche e governo del territorio. In: Parchi, no. 29/2000
14. Dematteis G (2011) Le grandi città italiane: società e territori da ricomporre. Consiglio italiano per le scienze sociali. Marsilio, Padua
15. ANCSA (1990) La nuova Carta di Gubbio. Un contributo italiano alla riqualificazione della città esistente, 11th National Meeting, Gubbio
16. Cattaneo C (1845) Industria e morale. Proceedings from the Society for Encouragement of Arts and Crafts, Milan
17. Gambino R (1983) Centralità e territorio, Celid, Turin

Acknowledgements

The researchers who appear in chapters 8.4, 8.5, and 8.6 have succinctly but rigorously responded to my request to present different areas of investigation related to research on landscape quality and quality of life. In particular, they have each identified the interactions between the different fields of research, with the aim of favoring complex systemic visions and adequate planning responses. Each contribution is organized into three parts: a description of the state of the field; defining the research by highlighting its advances with respect to the current debate; and the expected results and their impact in increasing quality of life. I thank them all: Vincenzo Riso (Minho University, Portugal), Paolo Castelnovi and Roberto Gambino (Polytechnic of Turin, Italy), Giorgio Osti (University of Trieste, Italy), Marco Frey and Massimo Battaglia (Scuola Superiore Sant'Anna, Pisa, Italy), Ilenia Pierantoni (Università degli Studi di Roma "La Sapienza," Rome, Italy), Riccardo Santolini (University of Urbino), Luca Cetara (European Academy of Bozen/Bolzano (EURAC) Research, Italy), Maria Teresa Idone, Roberta Caprodossi, Paolo Santarelli, Roberto Canullo, Giandiego Campetella, Stefano Chelli, Marco Cervellini, Renato De Leone, Rosalba D'Onofrio, Federica Ottone, Roberta Cocci Grifoni, Barbara Fenni, Sara Spuntarelli, Carlo Bisci, Bernardino Gentili, Alessio Acciarri, Gino Cantalamessa, and Giorgio Di Pancrazio (University of Camerino (UNICAM)), Beatrice Marinelli and Piersebastiano Ferranti (Terre.it, a UNICAM spin-off). Moreover, I thank Roberta Caprodossi, Daniele Fabiani, Corrado Gamberoni, Ilenia Pierantoni, and Andrea Renzi who have collaborated in the collection and insertion of images in the text. In particular, I thank Federica Sargolini with whom I have thought at length about the relationship between man and nature.

Introduction

The redefinition of happiness as "frugal abundance in a supportive city" is a recent oxymoron introduced by Serge Latouche[1] that is probably even more strident than "happy degrowth" or "sustainable development." However, this forces us to reflect. In a recent essay in preparation for the "festival*filosofia*" in Modena Carpi Sassuolo, Italy, the philosopher reminds us that we live in a world with a triple absence of **limits** [1]: in the production and therefore the withdrawal of renewable and **nonrenewable resources**; in the creation of **needs** and therefore superfluous products and waste; and in the output of slag and pollution (air, land, and water). We live in a society of "growth as an end in itself," where the dependence on consumption and the compulsiveness of desire do not lead to well-being. Latouche therefore poses the question regarding the sense of richness, bringing into play the "imaginary foundations" of the economy, and believing in the possibility of exiting the infernal circle of the "*unlimited creation of needs and products and the growing frustration it generates*".

The origins of this reflection in the contemporary political debate are ascribable to Robert Kennedy more than 40 years ago at the University of Kansas[2], where he gave a celebrated discourse on the gross domestic product. With simple words and great media effect, he demonstrated why an **indicator of richness** could measure only what industry produces, leaving behind the **quality of life** and immaterial heritage such as creative intellect or interpersonal relationships:

[1] For more information, see Latouche S (2012) Facciamo economia. In: La Repubblica, 14 September 2012, in which he criticizes the consumerist abundance that expects to generate happiness through the satisfaction of everyone's desires. Countering this logic, the society of degrowth proposes to make humanity happy through self-limitation, thus being able to reach "frugal abundance".

[2] On 18 March 1968, at the University of Kansas, Robert Kennedy brought up for discussion for the first time the GDP as an index to measure the richness and well-being of a population.

... we will find neither national purpose nor personal satisfaction in a mere continuation of economic progress, in an endless amassing of worldly goods. We cannot measure national spirit by the Dow Jones Average, nor national achievement by the Gross National Product ... It measures everything, in short, except that which makes life worthwhile, and it can tell us everything about America—except whether we are proud to be Americans.[3]

When this extraordinary speech by the American presidential candidate is cited, one often forgets that he would be assassinated just 3 months later in a hotel in Los Angeles, California.

The relationships between economic potentates, in the diverse ways they present themselves, and the way we live on this planet are cogent; being able to elude them is, perhaps, only pure illusion. This has pushed classical economics to consider Nicholas Georgescu-Roegen's [2] doctrine of **happy degrowth** to be irreconcilable with the concept of sustainable development, which is universally accepted, at least in theory [3, 4]. On the other hand, Serge Latouche, in his disapproving of the concept of rational and efficient development, does not omit criticism of the current concept of **sustainable development**, his opposition being directed at any increase in resource consumption [5].

Different movements of European and global thought, both governmental and not [6], and elaborators of strategies for the protection of the planet have also progressively pushed theories for the **conservation of the planet** toward those for **economic development**[4] [7–11]. Studies on nature and human activities have been opened to reciprocal contamination [12–14]. In reality, however, many of the changes that have recently arisen in the economy remain anchored to mutations and conditioning internal to the system. For example, as argued in the 1960s by David Harvey [15] in *Occidente*, the movement from a Fordist model to a post-Fordist model has produced extreme flexibility in the production system, which is capable of integrating and implementing very different modes, times, and places of realization in the network. However, this increased **flexibility** [16] has not signaled a real change in epoch or economic paradigm, but rather a mutation completely internal to capitalism. This has served exclusively to diminish the cost of labor, reducing the time between investment and realization of profit.

[3] A substantially new economic system, which many already call "cognitive capitalism," points at different accumulation processes and attributes important growth to the value of knowledge, affection, and relationships, both imaginary and symbolic. For more information, see Fumagalli A (2007) Bioeconomia e capitalismo cognitivo, verso un nuovo paradigma di accumulazione. Carocci Editore, Rome.

[4] From the Club of Rome in the 1960s to the European Landscape Convention (ELC) in 2000, great steps forward have been made. For more information, see Niccolini F, Morandi F, Sargolini M (Eds) Parks and territory. New perspectives and strategies. List-Actar, Trento-Barcelona.

The **planetary crisis** we are experiencing deals with ecological, social, and economic balances simultaneously. These last seem the most difficult to align, obfuscated as they are by the convulsive pulses of the markets. One often loses sight of the concept, brilliantly expressed by Amartya Sen, that market economics "*is not a self-regulating process. It needs humans' rational intervention*" [17]. Looking at medium-to-long range social and ecological horizons may be of use. However, it is useless to think about facing the triple horn of the crisis without bringing into play convictions, habits, procedures, governing models, and *techne*, i.e. humans' capacity to knowingly manage their relationship with nature. Humans' role regarding the changing of the Earth is not secondary, since their responsibility in the transformation and **consumption of resources** is so evident. Starting from ***techne*** is therefore essential for satisfying that aspiration to security regarding the future of the planet and improvement in quality of life, for which humanity profoundly feels the need.

This volume wishes to expand on these themes starting with the city, where more than half the Earth's population lives, gathering its close relationship with the contexts and acting on the quality of landscapes. In fact, the landscape is an important part of quality of life for people everywhere: "*both in urban areas and in the countryside, in degraded and high-quality areas, in areas recognised as being of outstanding beauty and in everyday areas*"[5]. Therefore, starting from the idea of landscape disseminated by the **European Landscape Convention** (ELC)—"*Landscape means an area, perceived by people, whose character is the result of the action and interaction of natural and/or human factors*" —we examine, in particular, the interactions among the different biotic and abiotic components that contribute to environmental and landscape quality.

With this general objective of regenerating consolidated cities and **urban sprawl**, green spaces and **environmental networks** assume a central role. In particular, environmental networks that—even following important European Union directives and when being handled by the Council of Europe—are spreading throughout Europe slowly but perceptibly, innervating agricultural and urbanized areas, and are becoming the Trojan horse for the introduction of nature in cities. ***Greening the cities*** starts from the environmental networks [18].

The networks enter like rivers, extending to the free areas they find, often because they are residual or abandoned [19]; in some cases they contribute to **reshaping the city**, becoming structures; in others, they heal parts that are sick or in an advanced state of decomposition. Wherever they enter, they increase the quality of life [20]. Landscape urbanism and ecological urbanism have

[5] Recommendation CM/Rec(2008)3 of the Committee of Ministers to member states on the guidelines for the implementation of the ELC, adopted by the Committee of Ministers on 6 February 2008 at the 1017th meeting of the Ministers' Deputies.

been working on these themes for a while, even though the themes may seem to fall outside these disciplines.

Environmental networks signal conceptual and spatial advancement, as well as functional advancement, with respect to the tradition of "green belts" and "greenways." These can play a primary role in measuring what leaves the merely biological sphere and tends to assume a complex meaning. Full consideration of landscape diversity (also in the biodiversity matrix over the course of long and never-concluded coevolutionary processes), attention for the richness and diffusion of the cultural heritage and historical relationship networks, and the awareness of the density of the socio-economic dynamics that mold the territory give life to systems of "biocultural" connections that can reinvent the relationship between city and nature.

The last chapter of the volume concentrates on a case study of the Italian Adriatic city. This is investigated in three large fields essential for managing quality of life fertilely, through environmental networks: "Distinctive and Pleasant", "Efficient and Nice", and "Clean and Healthy". The research in progress is carried out through close interdisciplinary interaction between researchers in apparently widely separated sectors: ecology, planning, botany, forestry, chemistry, biology, geology, physics, sociology, economics, architecture, and engineering. They act through two lenses: **energy** and landscape, which together dismantle and rebuild the multiple levels of reading and likewise different problems on which all disciplines that affect our frameworks of life reason and work. The expected results are not completed, definitive responses regarding sustainable design, but rather structural references and orientations for planning and **decision support systems**, whose sharing within the **plans** and projects will favor:

- Greater sustainability of planned transformations and of those to be planned as a result of their **transdisciplinary** evaluation.
- **Inclusiveness** and implementation of policies and interventions of territorial worth with the goal of guaranteeing the economic, environmental, and social sustainability transformation interventions directed at the development of local communities.
- **Cooperation** between stakeholders and actors who govern the territory and who feel the need for an tool that can permanently monitor complex environmental systems.
- Improvement of landscape quality, and with it, the quality of life of its inhabitants.

Considering that the scientific debate on these themes has generally stopped at a few trivial simplifications of interpretational and decision-making processes, which are almost always partial and unilateral, the openness to **systemic complexity** can provoke disorientation and bring out problems, unsolvable at first glance, such as opportunities for connections between **disciplinary sectors** and approaches that do not usually communicate with one another, and the need to find ways to confront synthetically objective evaluations and subjective evaluations linked to feelings, emotions, and **perceptions**. However, a deep reflection on these themes should begin immediately so as to favor new experimentation. The profound crisis in **traditional urban planning and design**, along with its incapacity to provide convincing answers to the demand expressed by vast social strata, is not determined "*by a lack of power, but rather by its cultural and technical unbelievability that has grown, not diminished, through recent experience*" [21]. The populace, finally, no longer seems willing to accept a poor quality of life, which unfortunately affects many European cities, and a courageous change of course is truly urgent. This book is a small first step in this direction; it certainly does not provide solutions, but quite possibly will open new paths of research.

References

1. Latouche S (2012) Facciamo economia. In: La Repubblica, 14 September 2012
2. Georgescu-Roegen N (2003) Bioeconomia. Verso un'economia ecologicamente e socialmente sostenibile. Bollati- Boringhieri, Turin
3. Latouche S (2004) Degrowth economics: why less should be much more. Le Monde Diplomatique
4. Latouche S (2006) Le pari de la décroissance. Fayard, Paris
5. Latouche S (2011) Come si esce dalla società dei consumi: corsi e percorsi della decrescita. Bollati Boringhieri, Turin
6. Worldwatch Institute (2007) Vital Signs 2007-2008. The Trends That Are Shaping Our Future. Norton
7. Brown L (2002) Eco-economy. Editori Riuniti
8. Costanza R, Graumlich L J, Steffen W (2007) Sustainability or Collapse ? An Integrated History and Future of People on Earth. The MIT Press
9. Eurostat(2001) Economy. Wide material flow accounts and derived indicators, a methodological guide. Eurostat
10. Folke C et al (2002) Resilience and Sustainable Development. Building adaptive capacity in a world of transformation. World Council for Science (ICSU)
11. Meadows D et al (2006) I nuovi limiti dello sviluppo. Mondadori
12. Gambino R (1997) Conservare innovare, Utet, Turin
13. Sargolini M (ed) (2005) Trasformare/conservare. Spazio Ricerca, n.5, University of Camerino
14. Sargolini M, Talia M (eds) (2009) Prevedere/Anticipare. Spazio Ricerca, n.5, University of Camerino
15. Harvey D (1989) The Condition of Postmodernity. Blackwell, Oxford
16. Bateson G (1972) Steps to an ecology of mind. Chandler Publishing Company, San Francisco
17. Allix G, Caramel L (2009) Non si vive di solo Pil, intervista a Amartya Sen. In: La Stampa il 10 June 2009
18. Tolley R (1990) The greening of urban transport: planning for walking and cycling in Western cities
19. Clement G (2005) Manifesto del terzo paesaggio. Quodlibet, Macerata
20. Sargolini M, Caprodossi R, Gamberoni C (2009) Aree dismesse e continuità ambientali. In: Barbieri P (ed) hyperadriatica. Opere pubbliche e città adriatica. Indirizzi per la riqualificazione dei progetti urbani e territoriali. LIST, Barcelona
21. Secchi B (1989) Un progetto per l'urbanistica, Einaudi. Turin

Contents

Bios, *Techne*, and *Logos*

1

Tomas Kačerauska[1] conducted an etymological study on a few keywords associated with globalization: *bios*, *techne*, and *logos*. He carried out this study in the field of existential phenomenology, and suggested that our body is not separable from the spiritual environment. ***Techne*** manages the creation of interactive components between body and spirit. In transposing the reflection to a more extensive, planetary field, biological life (***bios***) is an inseparable part of the entire space–time system and is immersed in the spiritual environment. Conceptual reflection and language (***logos***) allow the context of the spiritual environment to be defined. The art of creation and also of continuous intervention activities on the planet keep this unavoidable intertwining in mind. Therefore, just as we can never imagine our body separate from the machines that surround us, it is likewise impossible to think of the Earth without the transformational activities of humans. *Techne* is the center of our reflections and applications (*logos*), while always maintaining close contact with *bios*.

The relevance of technique in the modern age is introduced by Emanuele Severino's reflection on our civilization at the end of the 1970s:

> *The contrast between **Marxism** and technology in Russia (and in the world) is destined to resolve itself in favor of **technology**, because it has on its side such arguments as to produce in the masses the conviction that technology can realize what the Marxist revolution has only been known to contemplate, that is, the liberation of humans from need* [1].

Severino discusses the motive for which religious and social problems become purely technical problems. Furthermore, the attention to voices of science and technique seems also to include poets and writers, who tend to align with these

[1] Vilnius Gediminas, Technical University, Department of Philosophy and Political Theory, Saulėtekio al. 11, LT–1023 V.

M. Sargolini, *Urban Landscapes*,
DOI: 10.1007/978-88-470-2880-7_1,

ideologies. Perhaps it is confusing to think that poetry, or more generally, artistic creation, agrees with technique. Doing so, how could "**creativity**" ever continue to "*be ahead of the times in which it creates*" [2]? If **technique** is the normal course of things, why should great creators, those who capture what others flee from, place themselves second? Moreover, would such an idea of technique also introduce less of any form of social or political meditation, or even "social revolution", in the **future** [1]? In fact, it would no longer be necessary to lavish anything on ourselves due to our conviction in a particular lifestyle, since technology would become the instrument for satisfying any **human need**. Thus, there is no need to convince someone regarding an attitude of life, since everything can be conceded and humans can, finally, concretely satisfy their own needs.

But what acceptance of "technique" are we talking about? According to Severino, it is not possible to understand the authentic sense of "technique" guided by modern science, if we do not go back to the oldest Western body of thought—Greek philosophy—and if, together, we are not able to sense the deep unity that links technique to philosophical thought in the last two centuries.

Technique in our time is the "*most radical form of techne*" [1]. The word *techne* expresses the way in which the Greeks thought about human actions; thought that, by its nature, was founded on "becoming," on the oscillation of things between being and nothing. The concept of the transformation of nature as the work of humans that we discuss today therefore has these profound roots. For **Heraclitus**, "**becoming**" is the substance of a being since each thing is subject to time and transformation. Even what seems static to our senses is, in reality, dynamic and changes continuously. This concept is realized in the thesis that identifies fire as the beginning of all things. Fire symbolizes movement, life, and destruction. "Becoming" is therefore the immutable law, *logos*, since "*everything mutates, except for the law of mutation*" [3], which regulates the alternation between birth and death. "Becoming" is composed of opposites that coexist within things.

Aristotle goes beyond Parmenides's noted oppositions to "becoming" ("*what is, is and cannot not be; the nonbeing is not and cannot, in any way, be*" [3]) and the so-called pluralist, materialist philosophers, such as Anaxagoras, who likened "becoming" to **being**, starting from the certainty, given as evidence, that sensing beings were continually subject to "becoming" (and therefore to corruption and death) and moving. Therefore, Aristotle occupied himself with movement—intended as the passage from a certain type of being to another certain type of being—time, and physical phenomena in general, providing one of the first complete studies of physics.

Among modern philosophers, anchorage to the Greek sense of becoming, with some adjustments, was pursued by Hegel and Bergson, and it is notably through Emanuele Severino that a return to Parmenides is made, in a dramatic countertrend [4]. At this point, we cannot dwell on anything other than the philosophy of the sense of being with its references to becoming or mutating. It is interesting, however, to highlight that *techne*, which represents the actions of humans, is not extraneous to this comparison.

More at the heart of the human/nature relationship, we instead find the "**cosmic pessimism**" of Giacomo Leopardi. In his *Dialogue between Nature and an Icelander*, the poet no longer considers Nature as the "benign mother" of living beings. The guilty party is no longer humans, who have voluntarily strayed from natural laws, but Nature, which is indifferent and incapable of procuring joy for her children. An essay by Laura Sanò stands out [5] when expanding on this relationship. Sanò reflects on Plato's famous dialogue ***Protagoras***, whose principal subjects are **Prometheus** and Epimetheus. Epimetheus has to distribute "natural gifts" to mortals, but on reaching **humans**, he realizes he has run out of natural gifts. Prometheus intervenes and decides to steal fire and give it to humans. Humans are therefore given the technical skills necessary for life, but not political virtue. Sanò states that:

> ***Techne*** *does not pertain to humans' natural skill set, it is not an innate or congenital presupposition, it does not reside in humans themselves; rather, it is to be considered as an external gift, bestowed thanks to a ruse and a divine sacrifice* [5].

Techne has certainly evolved from Prometheus's gift to today, but the sense of a "non-natural gift" remains, and serves humans to become the masters of things and the authors of nature's transformations through efficient technological organization linked to the development of natural science. However, Maria Teresa Costa's hope [6], when referring to Benjamin's idea of *techne*, is that it can stream from the pure pleasure of *mimesis* and not become an end in itself. In such a sense, it should be capable of fertilely intersecting the Cartesian dualism that divides reality into ***res cogitans*** and ***res extensa***. *Techne* dialogues, on the one hand, with the nonextension, awareness, and liberty of the mind, and on the other with the extension, limitation, and unawareness of space. It undertakes in the territory that task of the "pineal gland" that Descartes imagined as a communication system in the human body between the two domains. "Creative people", the poets, and the artists referred to by Ruffilli, may also be interested in this type of technique [2].

References

1. Severino E (2002) Téchne. Le radici della violenza. BUR Rizzoli, Milan
2. Ruffilli P (2011) Il resto dei tempi. Prisma economia. Società, lavoro. IRES Marche. n. 1, Franco Angeli, Milan
3. Reale G (2006) I presocratici. Testimonianze e frammenti. Bompiani, Milano
4. Severino E (1964) Ritornare a Parmenide. Rivista di filosofia neoscolastica
5. Sanò L (2007) La scommessa di Prometeo. Fra Leopardi e Nietzsche. In: Sanò L (ed) Le potenze del filosofare. Lógos, téchne, pólemos. Il Poligrafo, Padua
6. Costa M (2007) Povertà di esperienza nell'epoca della riproducibilità tecnica. In: Sanò L (ed) Le potenze del filosofare. Lógos, téchne, pólemos. Il Poligrafo, Padua

2 Ecology vs Aesthetics

"**Beauty** will save the world". This celebrated claim by Dostoyevsky uniting aesthetic and ethical aspects finds its roots in the Greek principle that beauty coincides with goodness [1]. In the Greek world, "beautiful" and "good" are so indissoluble that a term exists to describe both concepts: ***kalokagathòs***, "**beautiful** and **good**".

The union between the aesthetic and ethical planes would in part only come with Christianity, since this Greek mentality becomes undone through the expression of the idea of Good (and therefore True) in two ways: an interior way, for which the subject is posed as "good," and an exterior way, for which the subject is qualified as "beautiful." Here, naturally, the concept of beauty is much wider than its everyday meaning: the Greeks inserted the characteristics of harmony, symmetry, and eurythmics (exact and proportional rhythm) in this group. Even when one speaks of physical beauty one deals with interior beauty that is presented in the exterior image (*eikone*); it is beauty that reveals the interior superior **order**. It is enough to think of the Greek heroes; Achilles, an example for all, is handsome on the exterior, but is equipped at the same time with excellent moral qualities.

2.1 *Kosmos* and *Techne*

Kosmos refers to an ordered, harmonious world, gifted with justice; this harmony has both metaphysical worth (the world is structured according to a mathematical order) and aesthetic and sensible worth (the world is beautiful precisely because it is ordered). It is the idea of justice that presents itself as the ideal form of beauty. As Aristotle says in his *Poetics*:

> *What is beautiful, whether an animal or any other thing composed of parts, should not only have its parts ordered and in their place, but also have a size that cannot be casual; beauty, in fact, is in the size and order of the parts* [2].

M. Sargolini, *Urban Landscapes*,
DOI: 10.1007/978-88-470-2880-7_2, © Springer-Verlag Italia 2013

The Greek acceptance of "beautiful" is clear from this: something that is beautiful is not only pleasing to the eyes and ears, but also involves the qualities of the character, so that "beautiful" can also be correct behaviors, or better still, the dignity that such actions carry with them. Beauty is also something good, because beauty reveals the structure of the being and therefore its goodness, since, as Plato taught, being and wellness are ultimately the same thing [3, 4].

Much later, Kant, in his ***Critique of Judgment***, would argue that beauty is not an objective quality (property) of things. Beautiful objects in themselves do not exist, but it is up to humans to attribute such characteristics to objects. Beauty is therefore subjective, in the sense that it is something attached to the subject and not to the object, according to Kant. **Aesthetic judgment** based on the sense of beauty is what we use to notice beauty and harmony in a work or a landscape, creating an agreement between the sensible object (what we perceive and what we "reflect" on) and the need for liberty (which we feel freely). For Kant, the feeling of beauty is: (1) pure, in that it is not linked to the real existence of the object represented; (2) disinterested, in that the beautiful object should not respond either to a utilitarian scope or to altruistic imperatives; (3) universal, in that beauty is what pleases universally, shared by all; and (4) necessary, evidently not for a logical need, since explicit rules are not required for aesthetic judgment.

The order given by beauty can be found in nature, as *kosmos*, but it can also be artificially reproduced through a **representation of reality** that reconstitutes a new order: it is what the artist does, not in the restricted sense that we mean today, but in the wider sense of Greek culture, expressible as *techne*, a broader concept. According to Aristotle, *techne* goes even beyond experience, in that it puts order in itself. We have seen how *techne* has something to do with the human activity of manual production that presupposes abilities and construction regulations, i.e. rules, and is often translated into the term "art." Art, beyond coinciding later with the so-called "**fine arts**," is also the manual art of any artisan, who can be dedicated to many different types of activity, reach perfection, and become a master, an ***architéktōn*** [5, 6]. Art, naturally, in that creation, or perhaps recreation, always assumes an imitation, a reproductive impulse, a ***mimesis***. *Techne* is therefore the technical ability, the manual work of the builder, one who constructs so as to recreate order, the artificial *kosmos* that becomes more beautiful the closer one approaches the natural, original *kosmos*. Art has beauty as a goal, but beauty that is also order and harmony, and is therefore also good.

The proximity between *techne* and *kosmos* is therefore well structured in Greek philosophy and passes through what is beautiful and good. With a jump of more than 2,000 years, this proximity is charged with sense and topical interest in the relationship between aesthetics and ecology, and becomes the heart of the reflection on the debate in support of new strategies for urban and territorial regeneration.

2.2 New Contacts Through the Landscape

The landscape is probably the right approach to re-establish vital contact between aesthetics and ecology. Managing the landscape requires, on the one hand, cooperation with energies put into play by nature [7, 8], which we are not used to manage, and which are the results of plant and animals, as well as solar, wind, and water sources; on the other hand, it requires attention to the "soul of places," that extraordinary heritage of urban and rural culture that sometimes risks being swallowed up by individualistic visions contingent on the use of the territory [9].

The notion of **landscape**, polysemous and complex by definition, expresses the sense of inclusiveness. It directs courses of different proximities, making explicit the relationship between object (i.e. the environment) and representation (i.e. its artistic, emotional, and intuitive perception), between naturalness and artificiality, between historicity and modernity, between reality and interpretation. Attention is directed both at the territorial characteristics (morphological, environmental, socio-economic) of a region, as a scientific concept originally developed in the **discipline of geography** in the 1800s, and in the way in which those same characteristics are perceived through individual senses and emotions, capable of being artistically communicated through figurative and verbal language, already present in the painting and literature of the Late Middle Ages. Unfortunately, up to the present time, the landscape reading has been based on two notably separate approaches:

1. The point of view of **architects**, who are dedicated to the formal search for compositional balances, looking at single building relationships and deluding themselves that an overall vision can be found *a posteriori*.
2. The point of view of **ecologists**, who are exclusively concentrated on the mitigation of impacts provoked by human activities to the detriment of essential resources, biocenoses, and natural bioconnectivity.

In the author's opinion, we are now reaping the bad fruits of a sterile separation between two schools that, in any case, represent two important ways of approaching the landscape:

1. From the **art of gardens**, which has opened the architectural path of the landscape, but which does not have the necessary background to manage the complexity of territorial transformations and therefore the planning of the landscape required by the European Landscape Convention (ELC)[1].

[1] The ELC is a treaty adopted by the Committee of Ministers of Culture and Environment of the Council of Europe on 19 July 2000, officially signed on 20 October 2000 in Florence. It was signed by 27 member states of the European Community and ratified by 10, including Italy, in 2006.

This path has its roots in the historical attention to open spaces (gardens and parks), with its explicit interpretations of nature regarding the work of humans and the desire to "*exceed or exalt their beauty, modifying the morphology of places, plant covering, and waterways, in an attempt to also limit their biological perfection in the light of new scientific knowledge*" [10]. In time, one has witnessed the movement from the idea of the landscape park, imprinted with the aesthetic search for romantic or natural inspiration, to the urban park, conceived to increase the quality of life of citizens through better "environmental hygiene" [11].

2. From **landscape ecology**, originally conceived as the interface between geography and ecology, developed as an interdisciplinary science in systems ecology. Landscape ecology integrates biophysical and analytical approaches with humanistic and holistic perspectives through the natural and **social sciences**[2]. Contact with the territorial question is lacking, and in some approaches suffers from a deterministic imposition of processes, as derived from the first theories by Ian McHarg [12]. Landscape ecology studies the distribution and the **form of the landscape** with the goal of understanding its structures, processes, and meanings. It constitutes an important reference discipline for many sectors applied to the planning and environmental fields, analyzing spatial and functional patterns and their respective changes in time regarding the landscape mosaic to define **ecological networks** or to study **biodiversity**. The weakness of this interpretation is seen in the lack of a formal verification of new ecological balances, considering that acting without a push in the direction of innovation is not coherent with the ELC, which expressly speaks of the "creation of new landscapes." It also lacks fertile interactions with **subjective interpretation**, collective sentiment, intuitions and images, and the necessary relationship with social and economic dynamics.

The new perspective of contact between ecology and aesthetics introduces **complex visions** capable of going beyond the design of gardens and parks, to investigate and evaluate the intimate connections of the environment and the landscape with the physical and spatial organization of the city and the territory, looking for allies in the urban plan, which should "*exploit local and regional characteristics and express the personality of the place and the region*" [13].

---

[2] Different authors that have delved into related themes present this young field with very different features. In particular, the interested reader may evaluate the following different approaches: Naveh Z, Lieberman AS (1984) Landscape ecology. Theory and application. Springer, New York; Ingegnoli V (1993) Fondamenti di ecologia del paesaggio. Cittàstudi Edizioni, Milan; Pignatti S (1994) Ecologia del paesaggio. UTET, Turin; Turner MG, Gardner RH (1994) Quantitative methods in landscape ecology. The analysis and interpretation of landscape heterogeneity. Springer, New York; Forman RTT (1995) Land mosaic. The ecology of landscapes and regions. Cambridge University Press, Cambridge; Farina A (2001) Ecologia del Paesaggio. UTET, Turin; Finke L (1993) Introduzione all'ecologia del paesaggio. Franco Angeli, Milan.

In Europe, the ELC becomes the political support for moving in this direction. In fact, each state adhering to the Convention should commit to integrating "*landscape into its regional and town planning policies and in its cultural, environmental, agricultural, social and economic policies, as well as in any other policies with possible direct or indirect impact on the landscape*"[3].

The contributions to landscape ecology, appropriately integrated with **aesthetic/perceptive evaluations**, feed the view of urban planning and contribute to the reconstruction of an interpretive and planning grammar [14], allowing the local culture to rethink itself and develop a "***conscience of place*** *in order to not lose itself pursuing the standardizing myths of economic globalization*" [15]. The planning interpretations in which aesthetic and ecological goals are epitomized try to respond to the needs of nature and culture. As the philosopher **Catucci** reminds us, beauty does not derive only from aesthetic artifice, but also from need: "*What has made a landscape, a park, an environmental oasis beautiful is not a determined choice of language or style, but the sense of need that we notice in each step of the project*" [16].

In short, beauty, true beauty, which finds its roots in **Greek philosophy**, will save the world, as Dostoyevsky's generous and candid Prince Myshkin proclaimed stentorianly; it is, however, necessary to look for it in the contacts between aesthetic and ecological perception, between the sublime and the necessary, to produce it with a "revolutionary" **urban planning project** [17], and support it politically, giving voice to the social demand for a pleasant environment, with particular attention to the "right to the city" [18].

References

1. Werner J (1945) Paideia, the ideals of greek culture. Oxford University Press, Oxford
2. Paduano G (1988) Aristotele. Poetica, traduzione e introduzione. Laterza, Bari
3. Friedlaender P (1979) Platone. Eidos-Paideia-Dialogos. La Nuova Italia, Florence
4. Vidya monthly periodical (2006) Agathon: il Bene incondizionato secondo Platone. Rome
5. Bodei R (1995) Le forme del bello. Il Mulino, Bologna
6. Sergio G (2008) Storia dell'estetica. Laterza, Bari
7. Clement G (2005) Manifesto del terzo paesaggio. Quodlibet, Macerata
8. Clement G (2006) Où en est l'herbe? Reflexions sur le jardin planétaire. Actes Sud, Arles
9. Castelnovi P (2000) Symposium *Il senso del paesaggio*, ISSU, Torino, 1998; published by IRES Torino
10. Calcagno Maniglio A (2008) Dai giardini, ai parchi, al paesaggio. In: Sargolini M (ed) La pianificazione delle aree protette. Urbanistica Quaderni. INU Edizioni, Roma
11. Jellicoe G A (1969) L'Architettura del paesaggio. Edizioni Comunità, Milan
12. McHarg I L (1969) Progettare con la natura. Muzzio, Padua
13. Geddes P (1970, orig. ed. 1915) Città in evoluzione. Il saggiatore, Milan
14. Forman R T T (2008) Urban regions: ecology and planning beyond the city. Cambridge University Press, Cambridge

[3] European Landscape Convention, Art. 5, last paragraph.

15. Magnaghi A (ed) (2007) Documento programmatico. Piano paesaggistico territoriale regionale, Regione Puglia, Ass.to Assetto del Territorio
16. Catucci S (2012) Postface. In: Morandi F, Niccolini F, Sargolini M (eds) Parks and territory. List-Actar, Trento-Barcelona
17. Maldonado T (1970) La speranza progettuale. Piccola Biblioteca Einaudi, Turin
18. Lefebvre H (1970) Il diritto alla città. Marsilio, Padua

3 Environmental and Landscape Quality

The winds of **urban renewal** blowing throughout Europe are signaling the overcoming of rigidity, determinism, and the traditional hierarchical structure of planning processes at different governmental levels. Conversely, the themes of subsidiarity, sharing, and cooperation assume ever more relevance and centrality and, as a consequence, increase the interest in a strategic approach and in evaluation processes [1]. These two new requests hold a central role in the modification of physical space, are finely intertwined, and support each other with reciprocal pressure.

3.1 Evaluation Processes

Within the evaluation system, the redefinition of two domains is revolutionizing the overall system of processes: **environmental quality** and **landscape quality**. In Europe, there are evident contradictions in defining these two areas of intervention for at least two reasons:

1. At different governmental levels the environment is linked to **EU directives** and the landscape is dealt with by the **Council of Europe**, and an internal imbalance permeates these two issue centers that produce, respectively, implementation modalities and political orientations.
2. In the evaluation systems used in Europe up to now, which are based on an ancient legacy that is slow to die, the landscape component is understood exclusively from the aesthetic and perceptive point of view and is even described as an internal component of the environmental domain. This has occurred also in consideration of the fact that the first forms of quality evaluation were produced exclusively for the environmental area, and it is only in the last decade that the Council of Europe has tried to stimulate assessment regarding the landscape through the European Landscape Convention (ELC).

M. Sargolini, *Urban Landscapes*,
DOI: 10.1007/978-88-470-2880-7_3, © Springer-Verlag Italia 2013

In the first case, with regard to the environmental question, two rather noted procedures come up: the **Environmental Impact Assessment** (EIA)[1], and the **Strategic Environmental Assessment** (SEA)[2]. The EIA arises from a 1976 French law "*loi relative à la protection de la nature*" (translation: law for the protection of nature), in which three different levels of evaluation were introduced: "*études d'environment, notices d'impact* et *études d'impact*" (translation: environmental studies, impact assessment, and impact studies)". In Europe, it would become a supporting administrative procedure for decision-making authorities, aimed at identifying, describing, and evaluating environmental impacts produced by activating a specific project. The SEA, instead, is a process aimed at integrating considerations of environmental nature into plans and programs, to improve the overall quality of decision-making. The progenitor of the strategic evaluation method was the US Department of Housing and Urban Development, which published the Area-Wide Impact Assessment Guidebook in 1981. **The Netherlands** were the first to ratify the directives, followed by Denmark, Finland, Sweden, and then the other European states.

One has a partial and limited view of the landscape within consolidated **evaluation systems** in the EU. It is evident that it can seem relatively difficult to give a quantitative value to the landscape in all its complexity, while it is possible to identify and measure one of its many aspects (perceptive, ecological, historical/cultural, and so on). Because of this, in the experiences examined[3], one sees interpretations directed at reducing the complexity of a single aspect, or holistic visions entrusted with an overall judgment on behalf of experts that comply, however, with completely subjective evaluation methods. Through the ELC, a new political project for the landscape has begun, which is aimed at:

1. Recognizing the landscape's complex meaning[4].
2. Understanding its importance to the entire territory[5].

---

[1] In 1985, the European Commission issued Directive 337/85/CEE "*on the assessment of the effects of certain public and private projects on the environment*".

[2] The European Directive on the SEA (2001/42/CE) required all member states in the European Union to ratify the directive in national regulations before 21 July 2004. Many member states began to implement the directive starting from the narrowest terms connected to territorial planning, to later extend the approach to all policies bearing on the environment. The SEA Directive is directly linked to the EIA and Habitat Directives, in addition to other directives (water, nitrates, waste, noise, air quality) that fix requirements for the implementation and assessment of plans/programs.

[3] Research to identify indicators for landscape evaluation (to be applied in the SEA) was undertaken by the Umbrian Observatory for biodiversity, rural landscape, and sustainable planning.

[4] Art.1 of the ELC provides a definition of landscape: "... *it is an area whose character is the result of the action and interaction of natural and/or human factors*".

[5] Art. 3 of the ELC: "... *this Convention applies to the entire territory of the Parties and covers natural, rural, urban, and peri-urban areas. It includes land, inland water, and marine areas. It concerns landscapes that might be considered outstanding as well as everyday or degraded landscapes*".

3. Favoring integration in **territorial planning and programming**[6].
4. Involving **communities** and **interested populations** in the process of identifying objectives for quality to be pursued[7].

The **Recommendations of 6 February 2008** by the Committee of Ministers of the Council of Europe (guidelines for the establishment of the ELC, Art. 1, item 10) makes way for the opportunity to create landscape **observatories**, either specific, or as part of a wide observation system. A wide range of activities are assigned to the observatories:

- Describing the state of landscapes at a specific time.
- Exchanging information on policies and experiences regarding protection, management, planning, and participation.
- Using and processing historical documents regarding the evolution of landscapes.
- Fine-tuning quantitative and qualitative **indicators** to evaluate the effectiveness of **landscape policies**.
- Producing data used to understand tendencies and the development of previsions and future **scenarios**.

One of the first European experiments, the Landscape Observatory of Catalonia, and some noninstitutionalized Italian experience in the regions of Abruzzo, Calabria, and Sardinia, view the "*observatories as places to meet, in which expert knowledge intersects diffuse and ordinary knowledge, bringing together scientists, technicians, administrators, and representatives of civil society*" [2]. It therefore considers not only social processes through which communities are manifested, but also specific values that the subjects and interested populations attribute to the landscape. It is evident that this represents a substantial difference with respect to the deterministic-type approach regarding the environmental question.

3.2 Redefining the Environment and the Landscape

These new methods of landscape evaluation bring up the definition of the idea of landscape and, with it, the idea of environment. Up to now, the two terms have encompassed a very wide spectrum of interpretations [3], rendering them "right for many uses," but provoking confusion and difficulty for comparison and interdisciplinary communication. In some cases, the temptation has been to refine the many possible interpretations of the **environment** and **landscape**:

[6] Art. 5 of the ELC: "*... integrating landscape into their regional, town planning, cultural, environmental, agricultural, social, and economic policies*".

[7] Art. 6, paragraph C of the ELC "*... to assess the landscapes thus identified, taking into account the particular values assigned to them by the interested parties and the population concerned*".

natural environment, urban environment, contemporary environment, historical environment [4]; historical landscape, contemporary landscape, natural landscape, urban landscape [5]. The definitions of environment have gone from wide-ranging and all-inclusive ones—"*the complex of social, cultural, and moral conditions in which a person is found, formed, and defined*"[8]—to narrow ones operating squarely in the ecological school, where the concept of environment is complementary to that of **living organism** — "*all of the biotic and abiotic factors that act on an organism, population, or ecological community and influence its survival and development*" [6][9]. Regarding the concept of landscape, the definition given by the Devoto–Oli Italian dictionary in 1971 is what is still used in some **environmental impact studies**: "*a portion of the territory considered from the perspective or descriptive point of view, for the most part with an emotional sense that is more or less associated with an evaluation of artistic or aesthetic order*"[10]. Similar definitions in German (***landschaft***) or French (***paysage***) do not avoid this classical scenic interpretation.

An interpretation by the geographer Franco Farinelli strikes at the intrinsic ambiguity in the concept of landscape: "*the things and their image*" [7]. The openness and ambiguity of the notion of landscape was taken up by Alberto Clementi when he identified the divergence of interpretations and judgments of value as the cause of different cognitive perspectives [8]. Roberto Gambino would amplify this concept, speaking about useful and fertile ambiguity that recalls the tension between subjectiveness and objectiveness and drives back the objectifying charms of the Earth sciences [9]. But the naturalist **Friedrich Wilhelm Heinrich Alexander von Humboldt** had already gathered in 1860 the positive tension that is triggered between subject and object:

> *To hug nature in all its sublime majesty, it is not enough to obey external phenomena; it is necessary to show how this echoes within humans and how, in virtue of this reflex, it sometimes populates the sooty camps of myths with weightless images, and sometimes develops the noble germ of art [10].*

Internalizing landscapes to then return them under an artistic form becomes a way of expressing *techne*. It is an experiential landscape practiced by many artists. For one celebrated painter of Italian landscapes, **Tullio Pericoli**[11],

[8] Definition of "ambiente", Dizionario (Dictionary) Devoto–Oli, p. 197.

[9] Other definitions of environment: "... *in ecology refers to both physical and biological factors affecting organisms*", Sherman Hollar (2011) Ecology. Rosen Education Service Publishing; "*those variables that (...) through their effect on the individual, affect the vital rates of the population*", Salvatore JA, Klemens JA (2008) Ecological fitting by phenotypically flexible genotypes: implications for species associations, community assembly and evolution. Ecology Letters 11:1123–1134.

[10] Definition of "landscape". Dizionario (Dictionary) Devoto–Oli.

[11] Tullio Pericoli is a famous Italian artist who is very attentive to the interpretation of agrarian landscapes. His preferred subjects are gentle slopes in which vineyards, cultivated fields, shrubs, forests, ditches, and small valleys alternate with the farms. It is as if every part is stitched to another.

landscapes and portraits are almost the same thing. They go beyond the concept of landscape as a vertical transmission belt of genetic identity that allows for the reproduction of local societies. **Boesio** would write that Pericoli, "*in the furrows of the plough that digs up a field, in the wrinkle that lines a face, is not looking for photographic truth, but the mystery of the soul*" [11]. This expression of opinion synthetically encompasses the sense of landscape and, with it, all the complexity in evaluating its quality. All of these "liquid" aspects of the landscape are not easily traceable in the code; however, they are internal to the interactive triangle highlighted by Morin between "***physical sphere***, ***biological sphere***, *and* ***anthropic/social sphere***" [12]. For example, is **identity** only an attribute of the city or also a parameter for its evaluation [13]? The idea is that in experiencing their own territories, human communities produce values that are at least partly free of environmental data; they avoid deterministic interpretations and they necessarily go beyond the strictly economic sphere. The landscape therefore deals with values that cannot be examined exclusively within the category of the Earth Sciences.

3.3 Is a Synthetic Evaluation of the Landscape Possible?

At the base of environmental quality there is a systemic vision that, through an algorithmic consideration, brings together the measurable evaluations of different components that the environment can be divided into[12]. **Landscape ecology** has been working on these approaches successfully for a long time[13]. The same cannot be said for landscape evaluation. Interesting research by Claudia Cassatella and Attilia Peano on this theme is centered on ways to bring together different sector interpretations and different indicators (related to ecology, social perception, aesthetics, history, traditions, and the socio-eco-

[12] For further information: Bagliani M, Dansero E (2011) Politiche per l'ambiente - Dalla natura al territorio. UTET, Turin; Guarniero G (1993) L'impatto ambientale. informazione, analisi, valutazione, decisione. Alinea, Florence. See also the contribution by Piersebastiano Ferranti in the last chapter of this book.

[13] For a closer look: Brandt J, Vejre H (eds) (2004) Multifunctional landscapes: theory, values and history. Advances in ecological science. WIT Press, Southampton; De Groot RS (2006) Function analysis and valuation as a tool to assess land use conflicts in planning for sustainable, multi-functional landscapes. Landscape Urban Plann 75 (3–4); Breuste HJ, Niemelä J, Snep RPH (2008) Applying landscape ecological principles in urban environments. Landscape Ecology 23:1139–1142; Young J et al (2005) Toward sustainable land use: identifying and managing the conflicts between human activities and biodiversity conservation in Europe. Biodiversity and Conservation 14:1641–1661; Haines-Young R (2000) Sustainable development and sustainable landscapes: defining a new paradigm for landscape ecology. Fennia 178:7–14; Antrop M (2004) Landscape change and the urbanization process in Europe. Landscape and Urban Planning 67:9–26; Santolini R (2012) The indicators for the ecological network for monitoring of durable development in the Alpine regions. Platform Ecological Network of the Alpine Convention and Italian Ministry for the Environment, Land and Sea.

nomic dimension) in a **synthetic vision** of the landscape [14]. From this course, the difficulty in reaching an overall final synthetic vision capable of bringing together objective and quantifiable evaluations with others that are more subjective is evident. And the initial question that the authors pose is still open: "*If the landscape as a whole is impossible to assess from a holistic point of view, can we break it down into simpler elements that can be analyzed and monitored?*" [14]

Other experiences of environmental and landscape quality evaluation have been examined, using different criteria calibrated to the area of application and on the basis of different objectives, with mostly partial framing [15–17]. The experiences examined touch on the following areas [18]: relationships between rural landscapes and **agricultural policies**[14]; relationships between **rural landscape** transformations and biodiversity[15]; and relationships between **settlement sprawl** and rural landscapes[16]. In all of the cases investigated, the selection of indicators proposes following (not always successfully) criteria that are simply defined, highly comprehensible, economic with respect to measurement procedures, efficient in correlating different themes, and which contain a low number of parameters. The scale of reference is naturally different according to the evaluation objective. When the application regards purely environmental questions, the **indicators** are generalizable and applicable to different **contexts**. In facing the landscape question, the indicators tend to change with changes in landscape characteristics and population appreciation. In both cases, the methods maintain a certain degree of flexibility based on the availability of data.

[14] For further information: Guiomar X (2010) Periurban agriculture and local authorities: which policies for which proximities? Example of local policies in the Ile de France. In: Lardon S (ed) Agricultural management in peri-urban areas, Felici Editore srl, Pisa; Sargolini M (2009) Paesaggi agrari e continuità ambientali. Proceedings from the conference "Verso una politica del paesaggio. Temi e buone pratiche a confronto", Umbria Region. Attached to no. 21 Architettura del paesaggio - Paysage, Milan; Piorr HP (2003) Environmental policy, agri-environmental indicators and landscape indicators, Agric Ecosyst Environ 98; Landsis EEIG et al (2002) Proposal on agri-environmental indicators PAIS. Project summary. http://web.ccdr-alg.pt/sids/indweb/imagen/docs_extra/Outrosdocs/PAIS.pdf

[15] Several studies on indicators for rural landscape quality have been made by the Umbrian Observatory for biodiversity, rural landscape, and sustainable planning (coordinator: Massimo Sargolini). In particular, the indicators dealt with have included: (1) the state of the rural landscape and (2) the implementation and effectiveness of plans and programs that can have important effects on the landscape.

[16] A few studies have been made by the Umbrian Observatory for biodiversity, rural landscape, and sustainable planning (coordinator: Bernardino Romano) to look at indicators regarding settlement interference on landscapes and ecosystems. The parameters indicated are based on the quantity and geographical consistency of demographic components and urban settlements. Using appropriate minimum areas of statistical reference, correlation models were defined between the weight of settlement aspects and the qualitative/quantitative connotations of biodiversity.

References

1. Mascarucci R (2008) Goal Congruence. Meltemi Editore, Rome
2. Gambino R (2010) Osservatori in rete. In: Il paesaggio è solo una convenzione? Gli osservatori e la Convenzione Europea del Paesaggio. TAU Editrice, Todi
3. Zagari F (2006) Questo è paesaggio. 48 definizioni. Mancuso editore, Roma
4. Pane R (1967) Attualità dell'ambiente antico. La Nuova Italia, Naples
5. Cullen G (1961) The Concise Townscape, The Architectural Press, London
6. The American Heritage Science (2005) Dictionary, Houghton Mifflin Company
7. Farinelli F (1981) Teoria del concetto geografico di paesaggio. In: Paesaggio: immagine e realtà, Electa, Milan
8. Clementi A (2002) Revisione di paesaggio. In: Clementi A (ed) Interpretazioni di paesaggio, Meltemi Editore, Rome
9. Gambino R (1994) Ambiguità feconda del paesaggio. In: Quaini M (ed) Paesaggi tra fattualità e finzione. Cacucci, Bari
10. Von Humboldt A (1860) Cosmos. Saggio di una descrizione fisica del mondo, Venezia
11. Boesio A (2012) Il mondo di Pericoli in un ritratto. In: La Repubblica, 30 March 2012
12. Morin E (1977) Il metodo: ordine, disordine, organizzazione. Feltrinelli, Milan
13. Lo Piccolo F (1995) Identità urbana. Materiali per un dibattito. Gangemi editore, Tivoli
14. Cassatella C, Peano A (2011) Landscape Indicators. Assessing and Monitoring Landscape Quality. Springer, Dordrecht
15. Nijkamp P, Perrels A (1994) sustainable cities in Europe. Earthscan. London.
16. Williams F, Burton E and Jenks M (2000) Achieving sustainableurban form. Routledge, London
17. Bagliani M, Dansero E (2011) Politiche per l'ambiente Dalla natura al territorio. UTET, Turin
18. Peano A (2006) Il futuro nel mondo rurale. Alinea, Firenze

The City Under Transformation

4

Different studies by the Massachusetts Institute of Technology regarding scenarios for our future demonstrate that it is not possible to pursue unlimited economic, material, and quantitative growth in a world that presents well-defined biophysical limits [1, 2]. The city is the one most responsible for this uncontrolled and uncontrollable growth, which leads to unlimited land consumption. The urban conversion of land has progressively increased everywhere in Europe [3]. Italy leads the classification. From 1960 to 2002, in some regions of low settlement energy, i.e. where substantially stable demographics subsist, land consumption has been higher than 500%. The city has undergone a profound change in its substance regarding changes in:

1. Economic systems, starting from the **second industrial revolution** and continuing to today [4], with effects on the landscape from agriculture to industry, on the infrastructure, on the functional organization of the living/working system, and on the juxtaposition of functions for living with those for production, commerce, and residual rural areas [5].
2. The **social structure of the population**, especially starting in the 1970s, with the progressive improvement of average living conditions, the growth in levels of education, and the growing flux of immigration, tourism development, and city users [6].
3. The system of public and/or **collective spaces**, **places of collective identification** for meeting and public representation, once limited to the city's political centers, progressively moved to peripheral or marginal areas [7].

But the city has also been transformed in its appearance[1]:

[1] In particular, transformations linked to two social and economic processes that have forcefully emerged in the last 20 years should be examined—the reorganization of commercial complexes and the insertion of new immigrant populations—considering the logic of changes introduced by them in the city and territorial organization and the limits of local and urban planning policies➔

M. Sargolini, *Urban Landscapes*,
DOI: 10.1007/978-88-470-2880-7_4,

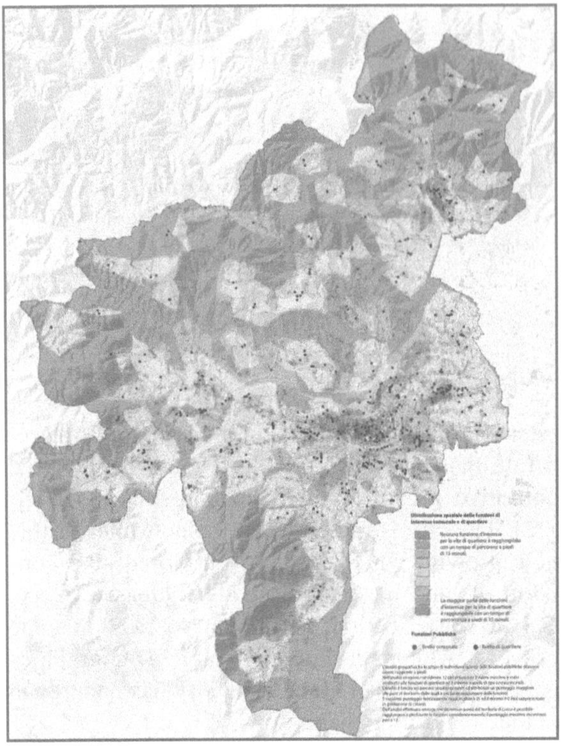

Fig. 4.1 Structural plan of Lucca, Italy. General hierarchy distribution functions: specialized functions of a district (neighborhood), municipal, and supramunicipal level are identified and scored

1. The system of **hierarchies**, **centrality**, and relationships between urban components, with the landscape moving from the centrality of place to the spaciousness of flux [8] (Fig. 4.1).
2. The relationships between the **historical city** and new architecture in **peri-urban areas** that tends to spread throughout the territorial context, losing contact with the center where it began [9] (Fig. 4.2).
3. The sign of physical city borders that are no longer readable, through unlimited urban expansion that registers the integration with other territorial and urban systems [10] and which makes identifying the "pertinent territory" always more difficult [11]. In this sense, marginal areas of the city in the 1980s and 1990s took on a fundamental role in transformation and **requalification policies** in European cities (Fig. 4.3).

instituted as of today. Some reflections on the theme can be found in: Lanzani A, Pasqui G (2011) L'Italia al futuro. Città e paesaggi, economie e società. Franco Angeli, Milan.

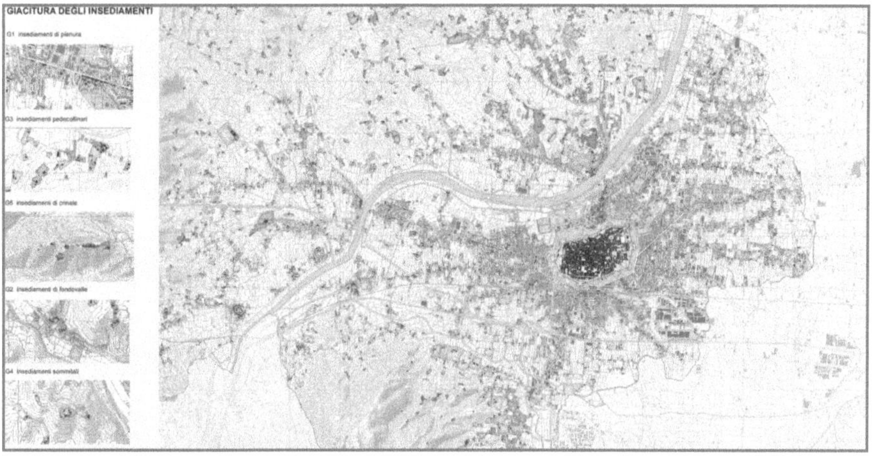

Fig. 4.2 Structural plan of Lucca, Italy. Morphological and figurative characterization of the urban fabric. The settlement pattern is analyzed on the basis of various factors and conditions: territorial, historical, functional, locational, and typological

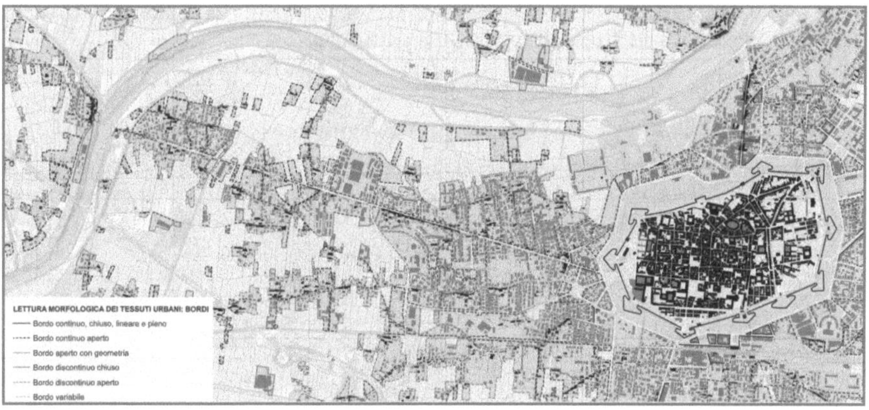

Fig. 4.3 Structural plan of Lucca, Italy. Arrangement of settlements. Urban edge analysis. Classification of the block's margin and general urban fabric construction and continuity of its linear and geometrical structure

4.1 The Diffuse City

These changes are particularly important in the Italian case studies that we present in this volume, but apply more extensively to the whole Mediterranean area [12], especially in some plain and valley floor areas and areas close to the sea (Figs. 4.4–4.6). The EU, in report EEA/2006, documents the proliferation

Fig. 4.4 San Benedetto del Tronto, Italy. Urban sprawl near the coastline

Fig. 4.5 Marche region, Italy. Urban sprawl of lowlands in the middle valley of the Chienti river

Fig. 4.6 Valle del Tronto, Italy. Urban sprawl in the valley

of consistent settlement diffusion in coastal areas that tend to join together various coastal centers, thus occupying the free areas. The original **multicentric model** tends to transform itself into a **metropolitan model**. To this phenomenon is added settlement diffusion toward rural hill areas in the near inlands, or hill centers rolling down into the valleys. In the case studies for the Marche and Tuscany regions, this tendency is very evident (Fig. 4.7).

The impression is that there is a growing attractiveness of some rural areas [13, 14] that become progressively occupied by new pseudorural settlements, tending, on the one hand, to increase the hemorrhage from historical, established, inhabited centers pouring out in a state of abandonment [15]; on the other hand, they tend to contribute to the formation of new intercommunity settlement systems, essentially implied cities.

In the literature, the imagination knows no bounds when introducing different definitions to describe the progressive changing of the city: from the "city-region" of Giancarlo De Carlo [16], to the "reticular structure" of Giuseppe Dematteis [17], or the "branching city" [18]; from "diffuse city" [19, 20] to "disperse city" [21], "infinite city" [10], or "dissolved city" [22] (Figs. 4.8–4.13). Finally, Pierre Donadieu's concept of "urban countryside" [23] describes a rural territory that can no longer be distinguished from the city. The common denominator in these different meanings is the accentuation of settlement dispersion that is manifest across the planet, with the appearance of great urbanized areas where the population is growing at the rate of 50 million per year.

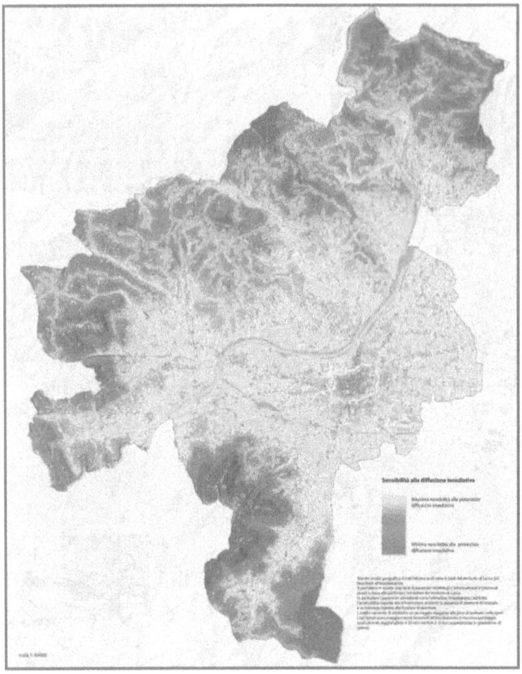

Fig. 4.7 Structural plan of Lucca, Italy. Tendency toward urban sprawl: more suitable urban areas for settlement growth are identified by geographic analysis. Morphological, infrastructural, and functional parameters are taken into account and "weighted" according to the particular features of the municipal area

Fig. 4.8 Athens, Greece. An unbounded urban settlement can be seen from the Parthenon

Fig. 4.9 Structural plan of Lucca, Italy. Output data rendering of the urban sprawl analysis defines the elements of interest and forms the structural design

Fig. 4.10 Umbria region, Italy. Rolling downhill from a ridge center that expands into the plain below

Fig. 4.11 Porto d'Ascoli, Martinsicuro, Italy. The dense city suffocates the "Sentina" Regional Natural Reserve

Fig. 4.12 Macerata, Italy. The dispersed city in the Chienti river valley

Fig. 4.13 Marche region, Italy. Small, rural settlements scattered in the hills and valleys

An epic phenomenon, according to the Population Division of the United Nations, which means only one thing: the Earth is at risk of coming up short in terms of environmental sustainability and **energetic consumption**. Not only that, population growth is confronted every day with the malfunctioning of the cities and their metabolism, through **settlement dispersion**, building waste, the consequent destruction of the **natural heritage** and **agricultural land**, traffic congestion, and finally, the catastrophic effects of climate change. All of these signals are manifest in the degradation of the landscape and the **quality of life** for city inhabitants, with a consequent profound crisis in contemporary living.

4.2 The City in Crisis

Instead of the many settlement models indicated previously, what about simply consolidating a model of "shrinking cities" [24]? Certainly, today a reflection is called for on how not to lead the city of tomorrow into collapse by activating a complicated lifesaving course in the name of sustainability, with the primary goal of improving the quality of life of the population.

In fact, in ordinary practice, urban and territorial planning has led to a reduction in the complexity of urban systems, considering them as substantially stable, believing that the transformations were preventable and controllable, and thinking about sustainability not as a constantly changing process and path that requires contributions from everyone, but as a final stable state. Interventions designed and programmed only on the small scale, for timely projects, which image the recomposition of a general mosaic after the fact, have been the main cause of the degradation we are witnessing. Unfortunately, responses have been of a sectoral, monodisciplinary type, with rather limited effectiveness, lacking a *modus operandi* that would allow possible alternative scenarios in policy, plan, and project choices for the city to be compared and their results to be monitored over time. The effects are for all to see: urban and peri-urban areas that are insecure and unprepared to face sudden climate changes; poor environmental quality, with air that is sometimes unbreathable, and a lack of policies to manage waste; energy-hungry urban areas that do not manage to limit consumption because of poorly designed urban systems and individual constructions and which are not able to start alternative energy production; areas devoid of meeting spaces for socializing, outdoor life, children's play areas, after-work hours for adults, and walks for the old. It is necessary to ask what has worked or what has worked only partially, and to be open to new challenges.

Some tools are used extensively and more courageously, for example, recourse to territorial urban adjustment [25], the development of the strategic and **evaluation approach** with argumentative and decisional goals. Some new suggestions have been introduced through European announcements, for example, "**smart cities and communities**"[2], and there are experimental projects already under way, such as "creative cities" [26–28]· However, very poor governing approaches linger when there is inadequate interpretation of the current change.

4.3 Interpretation and Representation of the Changes

Interpretation of the changes is very often entrusted to institutions, to general synthetic frameworks, without analyzing in detail the urban changes that are occurring in the **systemic network**. Just to give an example, in the case study of the **diffuse Adriatic city**, which is widely discussed in the last chapter of this volume, the current organization of the urbanized area, while not respond-

[2] See the announcement for the presentation of project ideas for "Smart cities and communities and social innovation" published by the Italian Ministry of Universities and Scientific Research with Directorial Decree N. 391/Ric of 5 July 2012, following Communication COM (2011) 808 by the Committee to the Council of Europe, the European Parliament, the European Economic and Social Committee, and the Committee of the Horizon 2020 Regions.

ing to a generally arranged project and lacking that formal and compositional balance we often evoke, effectively responds to some fundamental relationships between places for work and places for living or trade[3]. All of this does not mean that it can be considered satisfactory, nor should it simply be filed away. It is a system that has been well composed, first by planning its revision, but not losing its present positive pragmatism. Even more disregarded as an everyday problem is the graphical representation of the transformative dynamics of the city. Good representation certainly favors better participation by city government representatives[4].

In the literature, there are numerous studies on interpreting the dynamics of the **diffuse city**[5]. One recent publication on a study area in the **Umbria region**[6] can be considered original in the way in which, by using a wide and diverse range of knowledge and skills, it introduces new planning interpretation methods for the landscape, encouraging the incessant comparison between **natural processes**, inherited **cultural values**, and new contemporary opportunities carried along by the fluxes [29]. The changes that have affected Umbria in recent years, with regard to the size and means of development, have been in some ways unforeseeable (Figs. 4.14–4.16). In such a perspective, the following elements acquire particular importance: interpretive and descriptive activities leading to a recognition of hierarchy; ever more difficult distinctions between center and periphery or between city and **countryside**; typological conceptualizations; and research from a particular perspective to decode the multitude of objects and subjects that "*we recognize today in the city*" [30].

In the course of defining some types of landscape environments where urban and rural areas meet, the research awaits a vision that first allows for recognition and then for a continuous ordered story between tradition and contemporary

[3] A well-known case is the functional organization of the Marche [region] footwear district. More information can be found in: Corradini F, Dini G (2009) Logistica su misura. Franco Angeli, Milan.

[4] For a closer examination, see: Carta M (2011) La rappresentazione nel progetto di territorio, Florence University Press; Gambino R (1999) Le rappresentazioni come scelte di valore. Report to the seminar "Rappresentanza e rappresentazione nella pianificazione territoriale." IUAV, Venice, 16 December 1999; Magnaghi A (1999) La ricerca sugli atlanti territoriali. Report to the Seminar "Rappresentanza e rappresentazione nella pianificazione territoriale". IUAV, Venice, 16 December 1999; Beigel F, Christou P (1997) Paisajes épicos. In: Nuevos Territorios. Nuevos Paisajes. List Actar, Trento-Barcelona; Berger A (2007) Designing the reclaimed landscape. Taylor & Francis, New York; Morales Yago FJ (2007) Agua y paysaje en Yecla. Ayuntamiento de Yecla; Solá Morales M (1979) La cultura della descrizione/The culture of description. In: Lotus International, no. 23, April 1979. Electa, Milan.

[5] Several volumes include an important bibliography, e.g. Barbieri P (ed) (2008) Opere pubbliche e città adriatica. Indirizzi per la qualificazione dei progetti urbani e territoriali. List Actar, Trento-Barcelona; Barbieri P (ed) Hyper Adriatica. Op2 Opere pubbliche e città adriatica. List Actar, Trento-Barcelona; Bianchetti C (2003) Abitare la città contemporanea. Skira, Milan.

[6] The research, coordinated by Massimo Sargolini, was carried out by a working group at the University of Camerino in 2009–2010, within the activities of the Laboratory for Rural Landscape and Biodiversity.

Fig. 4.14 Umbria region, Italy. The plain with infrastructure and new residential areas of expansion

Fig. 4.15 Umbria region, Italy. The plain with infrastructure and new areas of industrial expansion

Fig. 4.16 Umbria region, Italy. Arterial fabrics in the plain

society to be reached. To identify harmony or contrasts between pre-existing and new forms of land use, 26 landscape contexts have been identified, testifying to the strong morphological distribution of Umbria (Fig. 4.17). In the different case studies, areas with wide panoramas alternate with closed, hidden, and remote areas; from general, all-inclusive reading to observations of individual items on the urban design scale, in which identifying elements and signs assume strong significance with respect to an overall glance and give strength to strongly cohesive places (Fig. 4.18). The research has lingered on this relationship on the basis of the following parameters:

- **Density**, keeping in mind that everything functions mostly by accumulation and continuous addition, not by composition (Fig. 4.19a, b).
- **Flexibility**, considering the kind of continuous oscillation in mobility that encompasses every urban planning event (Fig. 4.20a, b).
- **Casuality**, which distinguishes the juxtaposition of functions and forms (Fig. 4.21a, b).
- **Alternation**, identifying land uses and consumption that are intensive and strong in some parts of the city and rarefied, indecisive, or nonexistent in others (Fig. 4.22a, b).
- **Differences**, always to be evaluated, since they carry identifying value (Fig. 4.23a, b).

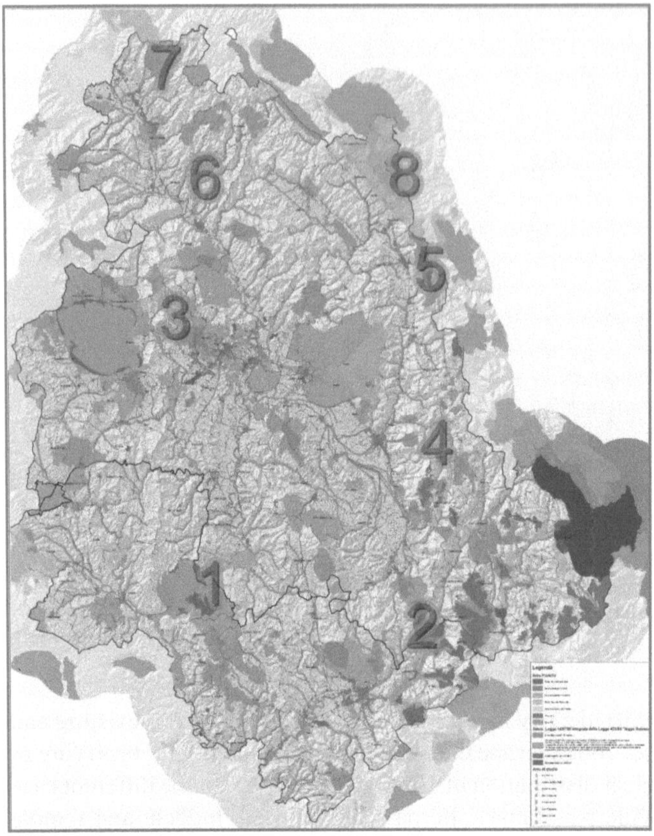

Fig. 4.17 Umbria region, Italy. Landscape areas: eight landscape study areas in close relationship with the natural protected areas have been identified

Fig. 4.18 Trevi, Umbria. The relationship between the historic city and the infrastructure of the plain situated below it

Fig. 4.19 a,b Monte Castello di Vibio and Faena Valley, Umbria. In the foreground, agricultural areas; and in the background, different degrees of accumulation

Fig. 4.20 a,b Faena Valley, Umbria. Historical Centre, recent expansions and environmental corridors

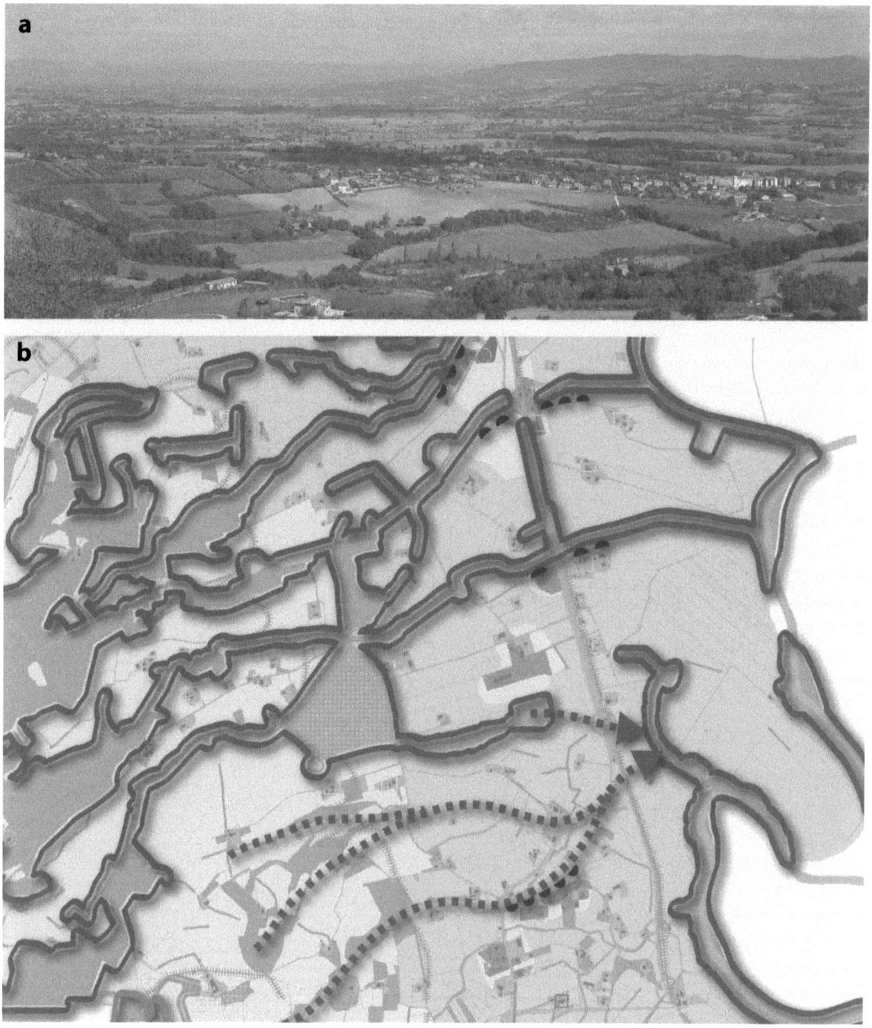

Fig. 4.21a,b Tiber Valley, Umbria. Random juxtaposition of different uses and destinations

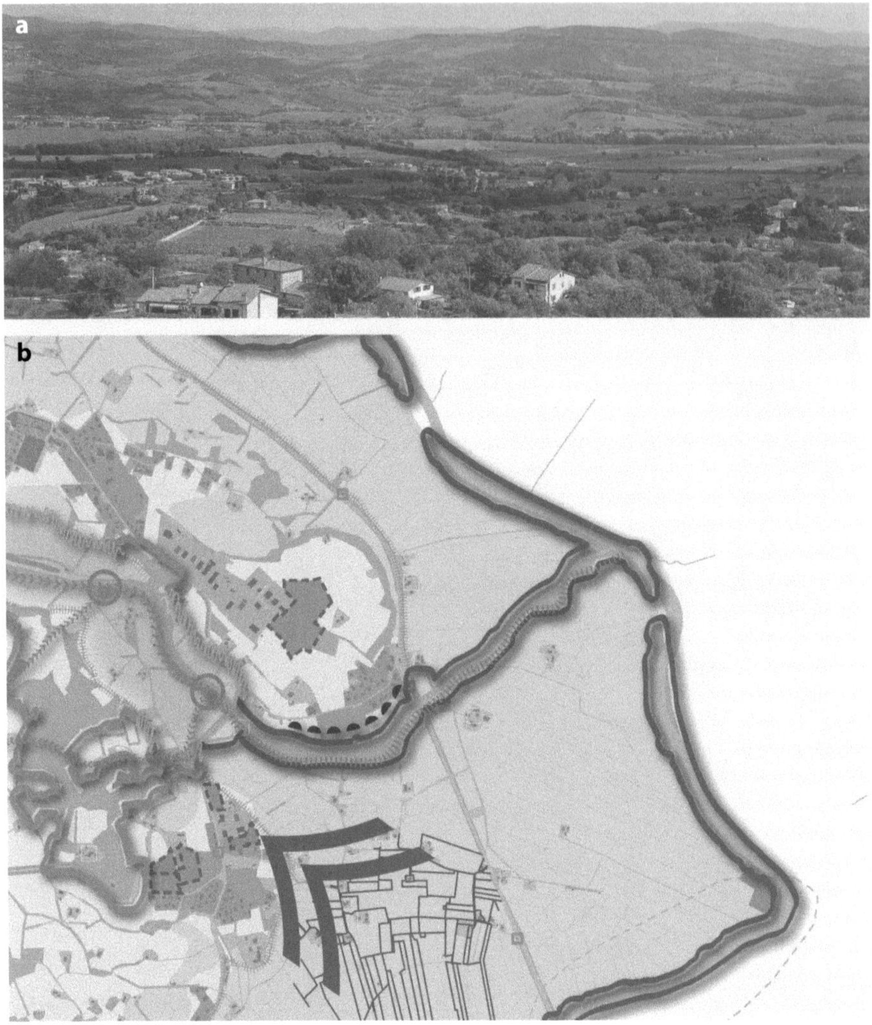

Fig. 4.22a,b Faena Valley, Umbria. The alternation between agricultural areas, houses, urban filaments, industrial and residential settlements

The originality of this recognition experimentation is, therefore, in the observation of the typifying characteristics, the new icons of transformation (from forms of growing renaturalization following abandonment and agrarian use of the land, to new intensive cultivation that tends to homogenize and make uniform the multiple varieties of the landscape; from "rolling into the valleys" from old settled areas on the hilltops, to the creation of large industrial slabs, to the construction of large commercial streets, to extensive suburban areas, to urban dust that spreads out into ex-rural areas, and so on), in their connection to spaces of **historical permanence** (historical centers, historical routes, isolated architectural properties, and so on). A

Fig. 4.23a,b Monte Castello di Vibio, Umbria. Historic center, residential area on the slope, green corridors connected to the river network

very articulated complex framework arises from this, which could become the basic layout for hypotheses of urban, environmental, and socio-economic regeneration on which the different levers of sustainability tend to operate (Figs. 4.24-4.29).

Photography has played an important role in this recognition, as shown by the photos in the present volume. In such a sense, the author would use photography as a tool providing us with contemporary art to understand the multicultural society in its relationship with living spaces and the environment, and to critically represent the landscape, seen as the privileged scene of the relationship between people, tradition, and change [31].

Fig. 4.24 Assisi, Umbria. The mingling between historic city and new urban areas

Fig. 4.25 Gubbio and the surrounding area, Umbria. The alternations of agricultural areas, houses, urban filaments, dense and compact urban settlement

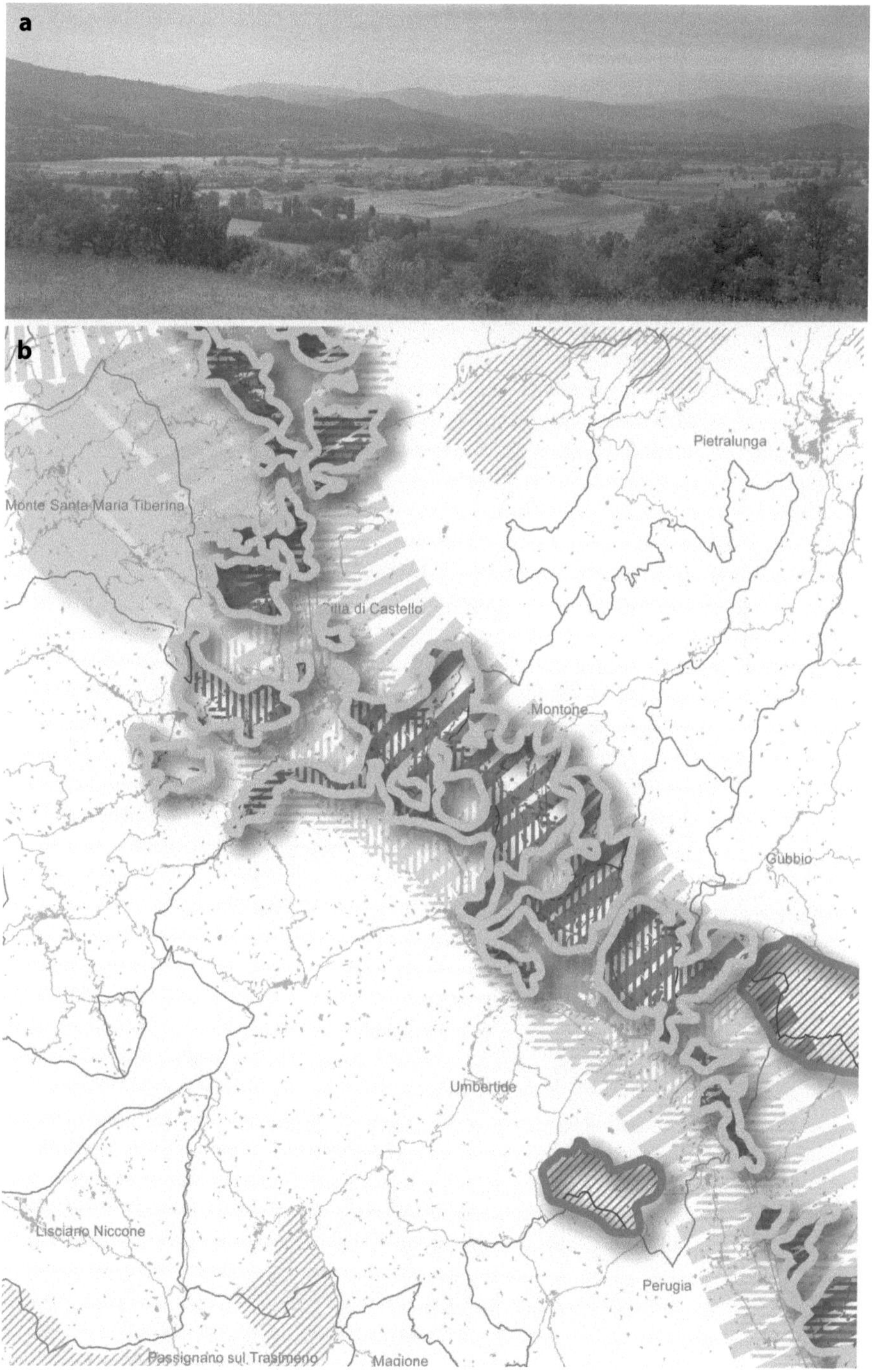

Fig. 4.26a,b Large industrial slabs in Città di Castello, Umbria. Intersections with the Umbria Regional Ecological Network

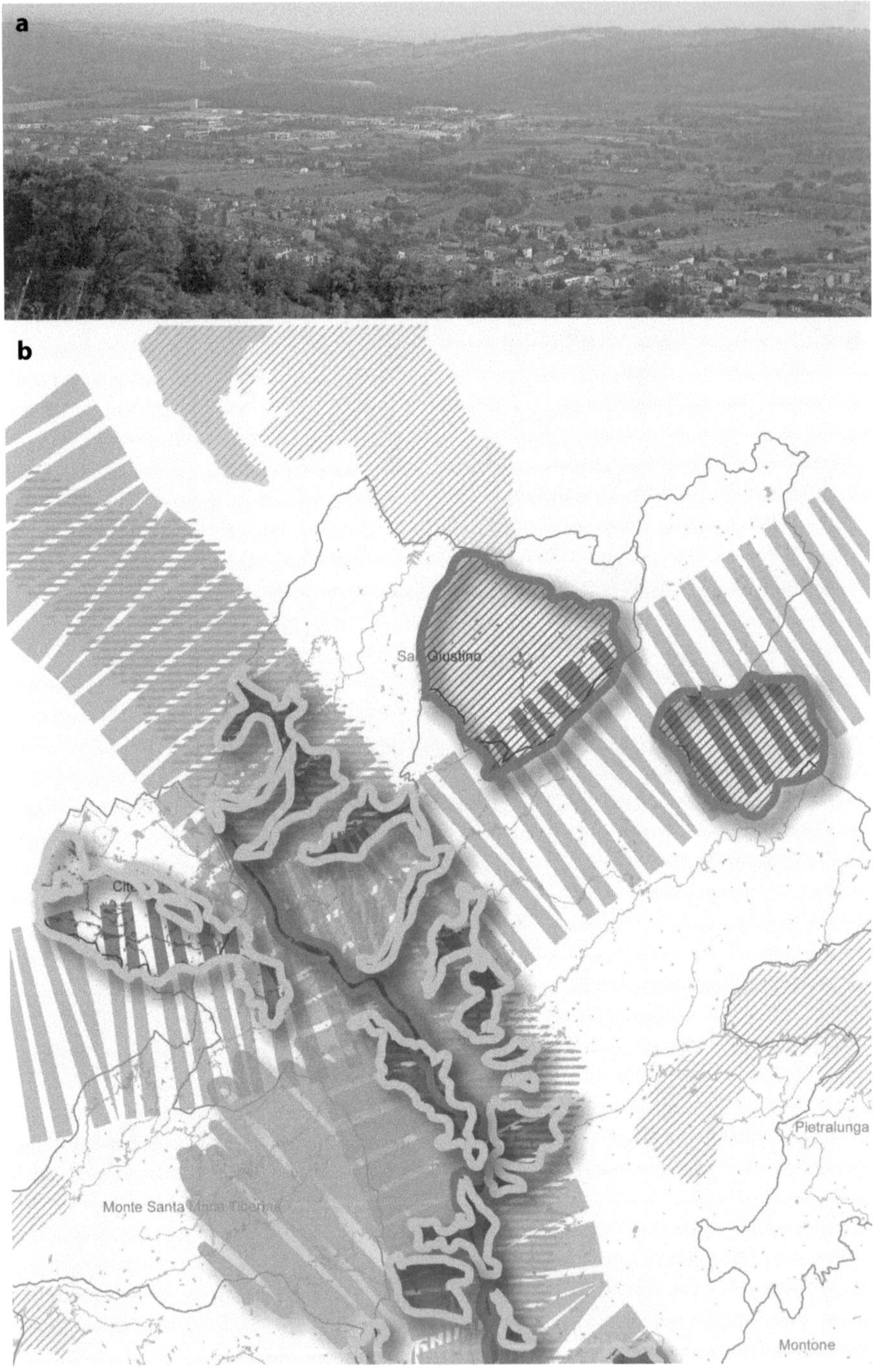

Fig. 4.27a,b A residential expansion in the surrounding area of Gubbio, Umbria. Intersections with the Umbria Regional Ecological Network

Fig. 4.28 Northern Trevi, Umbria. Infrastructural Intersections: fast tracks and service infrastructures for the agricultural activities

Fig. 4.29 Eastern Narni, Umbria. Important area of production in an agricultural context

References

1. Meadows D et al (1972) I limiti dello sviluppo. Mondadori, Milan
2. Meadows D et al (2006) I nuovi limiti dello sviluppo. Mondadori, Milan
3. Hamelin E, Razemon O (2012). La tentation du bitume - Où s'arrêtera l'étalement urbain? Editions Rue de l'Echiquier, Paris
4. Bonomi A, Abruzzese A (eds) (2004) La città infinita. Bruno Mondadori, Milan
5. Sargolini M (ed) (2005) Paesaggio territorio del dialogo. Kappa Edizioni, Rome
6. Vitali G (2012) Città per i cittadini, prestazioni territoriali e dotazioni urbanistiche. In: Sargolini M, Talia M (eds) Ri-conoscere e ri-progettare la città contemporanea. Franco Angeli, Milan
7. Sargolini M (2000) Paesaggi Insediativi. I luoghi dell'identificazione collettiva: problemi, contesti, differenze e relazioni. Urbanistica Quaderni, 24. INU Edizioni, Roma
8. Castells M (2003) La nascita della società in rete. Università Bocconi, Milan
9. Gambino R (1983) Centralità e territorio. Celid, Turin
10. Bonomi A, Abruzzese A (eds) (2004) La città infinita. Bruno Mondadori, Milan
11. Gabellini P (2001) Tecniche Urbanistiche. Carocci, Rome
12. Catalán B, Saurí D, Serra P (2008) Urban sprawl in the Mediterranean? Patterns of growth and change in the Barcelona Metropolitan Region 1993–2000. Elsevier, Amsterdam
13. Merlo V (1991) Il rurale uno spazio attrattivo? In: Sociologia urbana e rurale, no. 13
14. Osti G (2006) Nuovi asceti, il Mulino Ricerca, Bologna
15. Roncayolo M (1990) La ville et ses territoires, Gallimard
16. De Carlo G (1962) Relazione del seminario "La Nuova Dimensione della Città - la Città Regione". Stresa 19–21 January 1962. Istituto Lombardo per gli studi economici e sociali
17. Dematteis G (1985) Le metafore della terra. Feltrinelli, Milan
18. Detragiache E (2001) Bank Fragility and International Capital Mobility. Review of International Economics, 9: 673–687. doi: 10.1111/1467-9396.00306
19. Indovina F (eds) (1990) La città diffusa. Daest-Iuav, Venice
20. Pavia R (1996) Figure e luoghi della città diffusa. In: Ricci M (ed) Figure della trasformazione. Pescara
21. Camagni R, Gibelli M C, Rigamonti P (2002) I costi collettivi della città dispersa. Alinea, Florence
22. Secchi B (2005) La città del ventesimo secolo, Editori Laterza, Bari
23. Donadieu P (2006, orig. ed. 1998) Campagne urbane. Donzelli, Rome
24. Talia M (2012) Nuovi percorsi di ricerca. In: Sargolini M and Talia M (eds) Ri-conoscere e ri-progettare la città contemporanea. Franco Angeli, Milan
25. Ferraresi G (2009) Produrre e scambiare valore territoriale. Alinea, Florence
26. Florida R (2005) Cities and the Creative Class. Routledge, London
27. Carta M (2012) Quel motore nelle città creative. In: Il Sole 24 Ore, 22 March 2012
28. Carta M (2012) Creative city 3.0 New scenarios and project
29. Clementi A (2009) Oltre città. Verso una nuova urbanità medio adriatica. In: Barbieri P (ed) Hyper Adriatica. Op2 Opere pubbliche e città adriatica. List-Actar, Trento-Barcelona
30. Merlini C (2009) Cose/Viste. Letture di territori. Maggioli Editori, Milan
31. Basilico G (2001) L.R. 19/98. La riqualificazione delle aree urbane in Emilia Romegna. Editrice Compositori, Bologna

The Metabolism of Changes

5

Changes in the city have occurred, and are still happening, in such a rapid and uncontrolled way that it seems as if there has not been time for adequate metabolism, either on behalf of governance or on behalf of the settlement communities.

With regard to the former, the results are evidently ascribable to the very poor quality of public contributions (in the form of spaces, relationships between centralities, services) in new **contemporary conurbations**. Just to give an example, new **urban areas** are progressively being transformed into **multicentric urban–territorial systems**, but the supply of services has never been adequate for this conformation of the city.

The consequences are even more disastrous with regard to the latter: the lack of general organizational frameworks and shared strategic visions raises an ever more imposing wall between a consistent part of building construction and decision-making processes entrusted with the participating democracy.

5.1 Do Human Communities Still Produce Landscapes?

It seems as if the facts contradict **Cattaneo** who, more than 150 years ago, made the universal statement: "*Human communities are constructing landscapes*" [1]. For the first time, one has the feeling that the relationships between **local communities** (with their dynamics and socio-economic trends) and the **territory** are not forming landscapes. It is probably necessary to ask an essential question: if the landscape is determined by the encounter between community and territory, why do we regard some current products of settlement sprawl in different European cities as not pertaining to the landscape? Is it not the result of the desire of those who spontaneously created it[1]? Do we

[1] In this sense, reference is made to the concept of "spontaneous conscience", i.e. the attitude of a subject operating by itself and using the internalized culture exclusively, without the aid of ➔

M. Sargolini, *Urban Landscapes*,
DOI: 10.1007/978-88-470-2880-7_5, © Springer-Verlag Italia 2013

judge it through an assessment related to beauty, or perhaps functionality, or rationality?

We immediately dispel any misunderstanding. Neither beauty nor functionality or rationality can be considered as discriminating factors. We are not evaluating the **quality of the landscape,** rather if a given territory can be considered a landscape and consequently if it was made by the local community, i.e. if it was metabolized. The European Landscape Convention (ELC) is very clear on this point (Article 1(a), definitions): "*Landscape means an area, as perceived by people, whose character is the result of the action and interaction of natural and/or human factors*." It then adds (Article 2, scope):

> *Subject to the provisions contained in Article 15, this Convention applies to the entire territory of the Parties and covers natural, rural, urban and peri-urban areas. It includes land, inland water, and marine areas. It concerns landscapes that might be considered outstanding as well as everyday or degraded landscapes.*[2]

The resolving element is therefore the perception on behalf of the **population**. A rather slippery argument is entered into. This **perception** assumes the awareness of a place, which will then be the basis for the local project resulting from a real "*local participating democracy*" [2]. The local project is a cultural act, an opportunity for social communication, emergency, dialogue, and the encounter/conflict of interests and different expectations. The territory is certainly the favored place for any integration effort. But it cannot in any way be considered an incoherent sum of single technical/administrative acts or programs and timely, self-referential projects that characterize the current lack of a comprehensive, consciously shared plan for the structure of the city [3]. The extemporaneous desires of an individual that decides to "make a home" in a certain way and in a certain place, uncritically identified and sometimes ignobly supported by a consenting administration, cannot be confused with the collective conscious **perception of places**, which assumes cultivation, **critical reflection**, dialogic comparison, and strategic sharing.

On the other hand, if territorial choices do not manage to represent a wide range of institutions, interested subjects, and components of civil society, it is difficult to have the awareness and perception of a place on behalf of the community. In this case, there is no metabolism of the changes and a landscape is not produced. As Paolo Castelnovi notes in this volume, the same perception of landscape that the **ELC** approves, intended as an attribute of the population, is exposed to a basic contradiction. Within the definition of

mediation by a designer, who instead chooses and operates critically. See, in particular: Caniggia G, Maffei GL (1979) Composizione architettonica e tipologia edilizia. Lettura dell'edilizia di base. Marsilio Editore, Venice.

[2] Council of Europe, European Landscape Convention, Art. 2

landscape ("... *a landscape means an area, as perceived by people* ..."), in fact, a personal subjective interpretation of landscape is improperly likened to the collective vision of the population as a whole. We know that this common good [4] rests on the thousands of interpretations that subjective perceptions generate, but which should lead to a collective vision ("... *by people* ..."). Otherwise there would be no landscape. The homogenization of different perceptions occurs in time and through different methods according to the anthropological environment in which it is located: for example, it is certainly easier to homogenize the society of a country than that of a metropolitan hinterland.

5.2 The Common Sense of the Landscape

The **common sense of the landscape** is a resource for the identifying **value of living**, if it derives from planning and political work. If, instead, that landscape is poorly integrated with the living community, processes should be triggered to render collective a sense that would otherwise remain completely subjective and generate very different and contrasting results. As **Eugenio Turri** reminds us, integration begins with collective sensitivity to the landscape: there are societies that almost do not notice the landscape, while others approach it attentively. Care for the landscape "*is the work of social actors that know how to watch and get a sense of humans' role as a territorial agent*" [5].

Therefore, planning, within the new vision and ideas introduced by the ELC, should assume all of its responsibilities. It is up to this to support planning and political work that favors a common sense of the landscape, which is rooted in the population. In fact, Giuseppe Dematteis [6, 7] reminds us that a poorly integrated, nonmetabolized landscape signals the presence of a disintegrated **living community**. However, common sense should be reached without provoking the sterilization of vital potential in the relationships between individual people and the territory. Planning is the tool through which metabolism passes; because of this, it is responsible for the conscience of the landscape in a community. This can favor the regulation of relationships between the expectations of the community and territory, providing the technical/scientific support for the activation of civil society in public politics through participatory decision-making processes, toward innovative forms of deliberative democracy.

Community involvement requires a concrete basis that planning finds difficult to offer. The reasoned interpretation of places and the effective critical representation[3], whether of new areas under transformation or those where

[3] See Chapter 8.4.1 by Vincenzo Riso (University of Minho) in this volume.

Fig. 5.1 Umbria, Italy. A suburban settlement in the peri-urban area of Perugia

recycling or **upcycling** is being considered, are essential: first, because of the **community's perception**, and second because of the production of a **quality territory**. Following this phase, however, should come the involvement of the population that is not the usual *laissez-passer* stolen at the moment the project, prepared elsewhere, is presented to the public, but one that requires a proactive participatory process. On the other hand, the loss of the inhabitants' sovereignty over material and symbolic forms reduces the territory to an amorphous support of works and functions. Certainly, all of this is very far from any form of landscape, even a degraded one, as defined by the ELC. Metabolism, therefore, presupposes knowing how to recognize, interpret, and cultivate the community on the transformational dynamics under way, and opening the comparison of interpretation and evaluation parameters through audits and forums, which involve the local parties in the decision-making processes.

Much research on **park and protected area planning** in Europe, and more recently on landscapes in Umbria, Italy, as mentioned previously, has been centered precisely on investigating the metabolism level of several new industrial, commercial, and residential expansions in the fragile landscape balance (and therefore ecological equilibrium) of the Umbrian territory. Certainly in this, as in other case studies regarding the questions of today, the theme of the relationship of the **growing city** (the site of profound transformation) to the old city, with its great **architectural heritage**, **urban spaces**, and meanings that the past has transmitted, predominates. It is a comparison between processes of urban growth that are uncontainable and, in certain ways, uncontrollable; and fear is growing for the dissolution of the city via means of disperse settlements, for which it can even sometimes be difficult to understand the role and sense (Fig. 5.1).

In Umbria, consolidated residual landscape themes coexist with new territorial organizations; the prestigious traditional architectural heritage of the

Umbrian territory flanks contemporary urban expansions; the quick routes of logistics, commerce, and finance face the slow trends of tourism, teaching, and recreational fruition. The research raises different questions regarding some synergies or conflicting combinations between old landscapes and places undergoing change. What is the relationship between the population and these new scenarios? One glimpses a region in search for contact between old architectural and landscape figures and settlement, infrastructure, and technological innovations. It is the task of the territorial, landscape, and environmental project to investigate, in accordance with the common feeling of the interested population, new correlations between urban pre-existence and its recent dispersion.

Another interesting experience is taking place in Italy through the project of the **Marche Regional Ecological Network (REM)**[4]. Particular attention is being placed on the contact between network and city [8], whether for ecological needs (the need to favor permeability), or for returning a sense, through environmental infrastructure, to the splintered city that is dispersed and lacking in strategic visions. There are many project themes for **territorial regeneration** that spring up in these new visions. In particular, in the Marche case study, some clichés, looking at governance choices linked to the development of the sea and mountains as recreational landscapes, are coming into play. In facing these themes, important interaction and cooperation with the **regional plan**[5] is being activated. It is once again through the landscape, containing the collective perception, that the project for ecological regeneration, and consequently the transformation of the Adriatic city, is metabolized.

References

1. Cattaneo C (1845) Industria e morale. In: Proceedings of the Società d'incoraggiamento d'arti e mestieri. Milan
2. Magnaghi A (2000) Il progetto locale. Verso la coscienza di luogo. Bollati Boringhieri, Turin
3. Mazza L (2008) Il contenuto strutturale del nuovo piano, Il nuovo Piano. In: Proceedings from the 26th National INU Congress, Ancona (published in Urbanistica Dossier, 111/ 2009)

[4] The University of Camerino (UNICAM) spin-off "Terre.it" and the Marche region have just completed a search for the design of the Marche REM (Rete Ecologica delle Marche). The interdisciplinary working group, coordinated by Massimo Sargolini, was composed of researchers from the Universities of Urbino and Macerata, the Marche Polytechnic University (Università Politecnica delle Marche), and the Polytechnic of Turin (Politecnico di Torino).

[5] The School of Architecture and Design (SAD), UNICAM, and the Marche Region are completing research on "Landscape and biodiversity", in support of the revision of the landscape plan for the Marche Region, adapting to the ELC and the Code of Cultural Heritage and Landscape (principal investigator: Massimo Sargolini).

4. Mattei V (2011) Beni comuni. Un manifesto. Editori Laterza
5. Turri E (2005) Il paesaggio come teatro. In: Sargolini M (ed) Paesaggio territorio del dialogo, Kappa Edizioni, Rome
6. Dematteis G (2000) Il senso comune del paesaggio come risorsa progettuale. In: Il senso del paesaggio. IRES, Turin
7. Castelnovi P (2004) Il paesaggio limite del piano e del progetto. In: Quaderni della Rivista on doctoral research in landscape planning, Florence
8. Sargolini M, Cinquini F, Perna P (2006) reti ecologiche e siti Natura 2000, Il caso studio delle Marche. Kappa Edizioni, Roma

Changes in Vision

6

In the last ten years, it appears that people dealing with urban and territorial planning and design are putting much effort into ensuring the future of the city:

- Starting with the **tenth edition of the International Architecture Exhibition** in Venice in 2006, when **Richard Burdett** updated us with qualitative and quantitative assessments on the health of major world cities, and therefore on the health of the planet.
- Through national competitions and European announcements that decidedly move in the direction of experimenting with new and more sustainable ways of living and producing in cities, combating waste, and creating new **settlement forms** and **technological solutions**.
- Carrying out important experiments in urban design, in which natural components enter the city, relate with artificial components, and originate stimulating unions between the transformative actions of humans and natural spaces, in agreement with the reminders of **Landscape Urbanism** [1].
- With the support of the Council of Europe, which, in 2000, through the launch of the European Landscape Convention, indissolubly connected the landscape to quality of life and therefore to ordinary and/or degraded areas of the city and peri-urban areas. In reality, however, already in 1992, European institutions within the framework of sustainable development showed an awareness of the decisive significance of policies for urban areas with the European Urban Charter. The Aalborg Charter (1994), the Urban Acquis of Rotterdam (2004), the Bristol Accord (2005), the Leipzig Charter (2007), and the *Manifesto for a new urbanity–European Urban Charter II* in 2008 would all follow. The great objectives established in the treaties of Maastricht and Lisbon—social cohesion, economic competitiveness, and environmental sustainability—find confirmation in ever more precise and recurring terms in themes linked to urban projects and manage-

M. Sargolini, *Urban Landscapes*,
DOI: 10.1007/978-88-470-2880-7_6, © Springer-Verlag Italia 2013

ment aimed at the creation of "*attractive places where people want to live and work both now and in the future*" (Bristol Accord): **public space**; architectural quality of the anthropic landscape (to be promoted both through the conservation of the historical heritage and through new realizations); urban planning imprinted with the formation of compact settlements characterized by a mix of functions; infrastructure for intermodal mobility (with particular attention to pedestrian, cycling, and public transport networks); the requalification of degraded neighborhoods with actions coordinated in the urban and social plans; the recovery of decommissioned areas; **social housing**; and **urban services**.

6.1 Evaluating and Monitoring the Sustainability of Governing Actions

The debate on the sustainable city and quality of life in the city is like a large arena, where heterogeneous approaches and contributions originating in different disciplines are compared without converging on a **common vision** and, as a consequence, without finding integrated analysis and project tools suitable for facing the challenge of sustainability. Since the Brundtland Report, rivers of ink have flowed through this theme, numerous world encounters between superpowers and great environmentalists have taken place, and spontaneous acts on behalf of local communities have been made, but the risk is that everything ends with rhetoric. The conviction regarding the need to concentrate governing actions toward sustainability is unanimous, though divergent, even substantially, in the implementation paths chosen; also in this case, the key is ***techne***.

A large part of the scientific literature on the sustainable city defines criteria and analysis models that reveal a vertical approach, oriented toward concentrating on specific disciplinary fields, even though the EU, through the "Thematic strategy on the urban environment"[1] (2006) and the "Common framework of cooperation for the sustainable development of the urban environment" (Quadro comunitario di cooperazione per lo sviluppo sostenibile dell'ambiente urbano)[2] even earlier, call for coordination with other interested environmental policies, keeping foremost in mind the trans-sectoral nature of all strategies for improving the urban environment.

[1] COM/2005/0718 took account of the trans-sectoral nature of questions pertaining to urban management; any strategy for improving the urban environment requires coordination with other related environmental policies, i.e. the fight against climate change (buildings that favor energy efficiency, urban transportation plans, and so on), the protection of nature and biodiversity (reduction of the proliferation of cities, recovery of abandoned industrial areas, and so on), quality of life and health (reduction of atmospheric and noise pollution, and so on), and the sustainable use of natural resources in addition to the prevention and recycling of waste.

[2] Gazzetta ufficiale delle Comunità europee, L191/1 (13 July 2001).

Moreover, European research programs are also widespread, in which the qualitative requirements of the sustainable city are identified and related to both the environmental and socio-economic dimensions[3]. The applications that can be associated with this type of vision are those that study the **ecosystem** in the city or the city as an ecosystem [2]. In other studies, the concept of **energy** is introduced in transformation processes in our cities [3, 4], territories, and landscapes [5], investigating places and constructing devices for novel project scenarios in which energy awareness can generate new attitudes and behaviors [6].

The relationship between ecological and social design has also been looked at [7]. Attempts to define sustainability in more dimensions have been produced by Roberto Camagni [8] and Gianfranco Bologna [9, 10]. Additional examinations have been calibrated on the city—intended as the place of complexity–through the **general theory of systems**, which allows for an understanding of the current city in its parts, relationships, and laws, while trying to define what constitutes the DNA of the **urban organism**[4]. These approaches are all characterized by an attempt to broach the confines between different disciplinary sectors, but not yet between scientific knowledge and other types of interpretation by "city users" (citizens, professionals, public administrators). Finally, some authors have stressed a transdisciplinary approach in their research based on the search for key elements in the quality of life in cities, to be obtained by confronting experts of different disciplines with political and social actors [11].

In the international arena, the availability of sustainability indicator systems has felt significant impulses from the research of the United Nations Commission on Sustainable Development, the World Bank, and the United Nations Human Settlements Programme. Other studies have looked at the local scale of sustainability[5]. Next to sets of indicators that look mostly at single aspects in the urban ecosystem, such as atmospheric pollution, mobility, building, the ecological requalification of green spaces, and land use containment, other aggregate sets have been identified, as well as indices that condense more indicators, for example, the **global warming potential** or the LEED 2009 for Neighborhood Development Rating System developed for the

[3] The author is referring to European research programs like ECO-City.

[4] For further information: McLoughlin JB (1973) Guida e controllo del processo di mutamento. La pianificazione fisica come controllo dei sistemi complessi. In: McLoughlin JB (ed) La pianificazione urbana e regionale. Marsilio, Padua; Papa R (1992) La città funzionale. In: Beguinot C and Cardarelli U (eds) Città Cablata e nuova architettura, CNR-I.Pi.G.E.T., DiPiST University of Naples "Federico II"; Papa R, Gargiulo C (1993) Caos e caos: la città come fenomeno in Proceedings from the International Congress for the 21st Century – An Encyclopedia and a Project, Napoli Hotel Vesuvio–Paris, Palais des Congrès, 22 June 1992, DiPiST University of Naples "Federico II" and IPiGET CNR; Papa R et al (1995) La città come sistema dinamicamente complesso. In: Bertuglia CS, Fuccella R, Sartorio GL (eds) La città come sistema complesso in crisi strutturale-strumenti e tecniche per il governo metropolitano. Giuffrè, Rome.

[5] See, in particular, the indicators used within the United Nations Local Agenda 21 Programme for Sustainable Development or the European Common Indicators.

Congress for the New Urbanism based in Chicago, USA, which proposed measures for land consumption in American cities.

In the last chapter of this volume, the case study of the "**diffuse Adriatic city**" is discussed, which, prompted by experience already gained in the field, aims to produce methodological advancement regarding:

1. The interpretation and evaluation of urban complexity through **transdisciplinary quality indicators** that develop mutual correlations and adaptability to the context of interest.
2. The construction of an integrated body of knowledge of the city aimed at urban governance, making use of those parameters that best interpret the policies and decision-making strategies that are developed in the different territories.

6.2 Initial Experiences in Urban Regeneration

In 1993, the American Society of Landscape Architects adopted a "**Declaration on Environment and Development**", which is still current. This stated that it is necessary to start from the landscape connecting different components if we want to preserve the Earth. New transverse approaches are therefore necessary to face the themes of urban regeneration from new angles, assaying the fertility of conceptual development that faces the study of complex systems. In these new visions, nature is clearly called to make a relevant contribution to improving the quality of life, which is obtained by redesigning the urban landscape. For some time now, important international schools have concentrated on studies and examinations of the relationship between ecology and urban design. One is the traditionalist "new urbanism" that was developed in the USA, starting from theories and projects developed by the Leon brothers.

Another is the environmental/technological theme, which is attributed to Sir Richard Rogers and Sir Peter Hall. The main concrete result is the United Kingdom **Urban Task Force** ("**Towards a Strong Urban Renaissance**", 2005), in which a national program for the sustainable transformation of urban areas was drawn up. In different works produced by this school of thought, intervening in the urban landscape represents an opportunity to solve problems such as bad environmental quality (with particular attention to essential resources, such as air and water), a high consumption of nonurbanized land, the absence of neighborhood identity, the lack of improvement in areas and buildings representative of the community, and the lack of public and **collective spaces**. In regeneration processes in European cities, to answer the rights of the city, the environment, and the home, the questions to be faced can be organized into three large groups:

- The themes of **parks** and open and **green spaces** (linear and point) in the city and the immediate **context**.

- The themes of overall organization of the city and its energy needs on different building scales.
- The themes of the **safety** and quality of principal determiners of the urban environment (air, water, and land) on different city organizational scales.

Innovative experiences that are being tried in different European cities, starting from guiding actions directed at reducing land consumption, invite us to adequately improve areas that are still free, encourage forms of permanence of traditional uses of agrarian land in the city, and introduce forms of sustainable mobility.

6.3 Toward Demolition and Reconstruction

To start **urban politics** on a path of adapting the existing fabric rather than occupying new land, it is probably necessary to implement national policies that call for a cultural jump by all of the stakeholders. In particular, it is appropriate for designers to go beyond a traditional approach and businesses to free themselves from the habitual logic aimed at always building new constructions. Together, they should confront the taboo of demolition and reconstruction. Scrapping, similar to what is done with cars and household appliances, should be considered [12]. Some initial steps have already been made. For example, from the legislative point of view, in France, where national policies for urban rebirth have been implemented, a national law has been enacted to establish the **National Agency for Urban Renewal** (Agence Nationale pour la Rénovation Urbaine). From the point of view of programs realized, different important results have been obtained [13] in France[6], the United Kingdom[7], The Netherlands[8], and Spain[9]. To guarantee the possibility of new interventions, not necessarily demolishing and reconstructing on the same site, one should go beyond the expropriation approach, placing compensation principles alongside prerequisite principles. The experience of the "**transfer development right**," in use for some time in the USA, seems to be very effective [14]; recourse to territorial urban planning adjustment has also become increasingly interesting in urban design in Italy [15].

[6] In 2000, in Paris, the project Ilot Caillié was created. It was completed in 2007. Ilot Caillié, comprising 423 unhealthy buildings, is situated in the northeast area of the French capital.

[7] In the United Kingdom, the project for the Hulme–Manchester neighborhood (located in the suburbs of Manchester) was created in the 1970s. It dealt with an area of about 110 ha with a population of 12,000 inhabitants. Four large structures in the shape of arcs and 13 buildings in a line containing a total of 5,000 apartments were demolished and substituted with other apartments or buildings for other uses.

[8] In the 1980s, in The Netherlands, the Zuidwijk, Rotterdam Program was created. The quarter is found in the suburbs, south of Rotterdam. It was built after the Second World War and was arranged into eight areas; most of the estate was public.

[9] In Spain, the Turó de la Peira (Barcelona) project was created and the estate was built in 1959 under a private initiative. The neighborhood is situated in the northern part of Barcelona. The estate extends over 47 ha with 4,166 apartments distributed among 16 buildings.

Regeneration interventions will allow a rethinking of the network of centrality and places of reference, with particular regard for the new extraordinary role of the **urban fringe**. Particular attention is placed on the **microscale**, the care for public spaces, and social and economic fragility. Plans and programs will rely on design indications and types, as well as norms and parameters, to guarantee the best use of **natural resources**, mitigate climate variations that are under way, and prevent environmental risks. In particular, the objectives will be reached through:

- Adequate ways to arrange external spaces, foreseeing appropriate indices of land permeability.
- Innovative building types able to improve energetic efficiency and the use of design parameters that consider solar radiation incident on the building.
- A certification system that is univocal and credible on the national level, and which can be applied to all existing buildings.

Urban regeneration is therefore also ecological regeneration and technological innovation. Very often, however, as Federica Ottone reminds us in Chapter 8.5.6 of this volume, the urgency of intervening is discouraged by the incapacity to begin reflecting on the economic implications that an eventual technological and architectural reconfiguration of the public and private building heritage carries, and its real benefits for the community.

This assumption of responsibility implies that concrete cases for experimentation should be identified to effectively establish the point at which the advantages of an energetic requalification action can be sustained by collective and public administrations, and eventually what interventions are the most sustainable. When faced with complex problems related to climatic changes, we should fight the temptation to oscillate between the rhetoric on greenhouse gases risks and the freeing capacity of bio-architecture. Between these two extremes there is intelligent organization in ecologically effective shapes of the existing city; there is a close interaction between strategies for the city's economic development and the natural and historical/cultural permanence that our dense, stratified European territories are so rich in. In the case study of the Adriatic city, which is discussed in the last chapter, it is revealed how in areas of settlement diffusion there is a serious need to be able to adopt complex regeneration policies capable of harmonizing settlement structures that are growing in a casual way without the adequate involvement of public spaces and green areas and without evaluating, preventatively, the increase in energy consumption and the unsatisfactory level of territorial performance that is thereby derived.

6.4 The Potential of "White Areas"

Urban regeneration should keep in mind the land resources available to produce a good land project[10], especially when it deals with "**white areas**" [16], i.e. with "*still non-expressed potential*"[11]. In this spirit, in the author's opinion, we should interpret the state of transformation under way in European cities, which has profoundly altered the original fragile territorial fabric, transforming it into a never-ending succession of individual creations, in which free spaces become residential areas apparently lacking any sense, or becoming "negated territories"[12]. In the peri-urban fringe, the agricultural landscape disappears and the urban one, diluting itself into the rural area, fails to take shape. **Cityscape** and **landscape** are untraceable.

The changes, both when they occur spontaneously and uncontrollably and when they are programmed and realized in conformity with existing urban legislation, have left on the territory:

> *A quantity of indecisive space that lacks function, on which it is difficult to impose a norm. This combination does not pertain either to the territory of shadow or of light. It is situated at the margins: where the forests fray, along roads and rivers, in corners forgotten by* ***cultivation****, where cars do not pass. It covers disperse surfaces of modest dimension, such as lost corners in a field, vast and unitary, like bogs, moors, and certain areas abandoned following recent decommissioning* [17].

At times they seem like scraps, shreds, fragments. These "white areas" can be found, in particular, in ex-industrial areas that are now decommissioned, residual agricultural areas within the knitting of the urban framework, degraded and often abandoned areas (Fig. 6.1). For now they seem insignificant, but they could be the "new pages" that the territory needs to reorganize and regenerate shapeless spreads of "urban jam." New operating scenarios, new lands, or "waste areas" are opened, which in reality already existed but which our gaze was not used to noticing:

[10] The "land project," already outlined in the 1980s by Bernardo Secchi, was proposed again 20 years later. Secchi B (1986) Progetto di suolo, Casabella no. 520–521; Secchi B (2006) Progetto di suolo 2 in Aymonino A and Mosco VP (eds). Skira, Milan.

[11] Cf.: Ignacy Sachs, "Futuro", Enciclopedia Einaudi, Turin.

[12] This is a term introduced in the drafting of the Caserta Territorial Coordination Plan. The provincial territory was divided into settlement, agricultural and natural, and "negated" territories. This deals with areas of decommissioned or underused urban areas, areas pertaining to the infrastructure that have become open-air dumps, abandoned agricultural areas marked by land movements and abusive dumping, active or abandoned quarries, dumps, or sites of waste transfer.

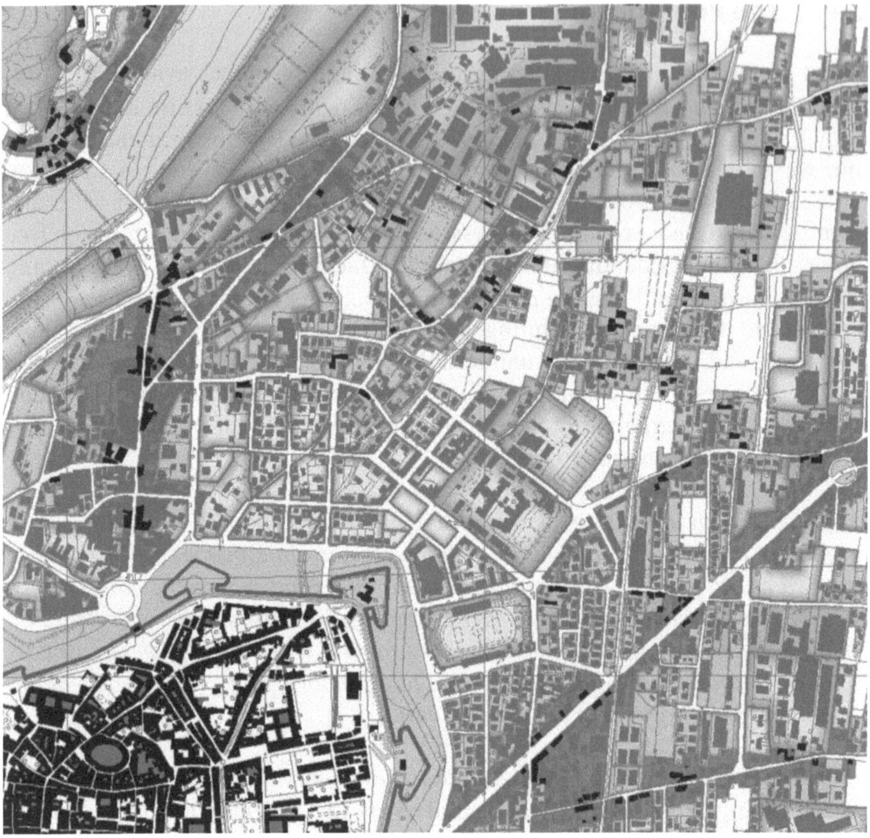

Fig. 6.1 Structural plan of Lucca, Italy. Historic city, new radial fabrics, and interstitial voids

> *Under this perspective, the possibility concretizes that the project can absorb the dynamics of time, creating strategies capable of annexing, selecting, or abandoning, and ultimately dialoguing with the waste rather than excluding it outright as not proper material* [18] (Fig. 6.2).

In research conducted on the Adriatic area in Italy, from Venice to Pescara, numerous **abandoned areas** have been identified following decommissioning, which has occurred often in recent times[13] (Fig. 6.3). They are either spaces of considerable dimension, or smaller, **diffuse spaces** that are almost invisible;

[13] The examination of decommissioned areas in the Adriatic city was realized under the research project *Programmi di Ricerca di Rilevante Interesse Nazionale* (Research Programmes of National Interest) PRIN 2006, "Public Works and the Adriatic City." For each area, the role that it could assume in the recomposition of the environmental structure of the city was identified on the basis of its specific location and in relation to the different systems of landscape in which it falls and its immediate territorial context.

Fig. 6.2 Bochum, Germany. An industrial area becomes an urban park (by Frank Vincentz; own work) [GFDL (http://www.gnu.org/copyleft/fdl.html) or CC-BY-SA-3 (http://creativecommons.org/licenses/by-sa/3.0)], via Wikimedia Commons

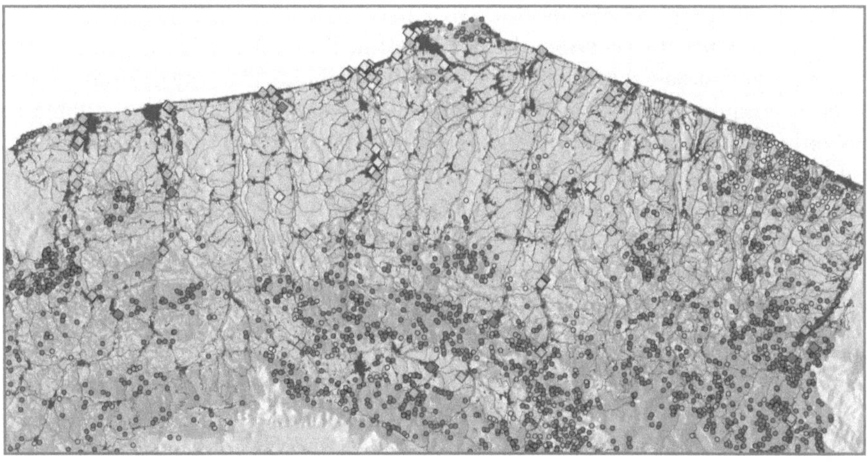

Fig. 6.3 Marche region, Italy. The decommissioned industrial areas, represented by square symbols, are particularly common in lower river valleys. Abandoned agricultural areas, represented by small circles, are common in the foothills of the Apennines and near the southern coast

the sparser the **city fabric**, the greater the number of smaller spaces. They can crop up in decommissioned production sites or abandoned agricultural areas; many of them are available for new uses. From these residual areas (which could take on the role of new landscape/environmental centrality functional for the urban machine), it is necessary to start again to reorganize the city, redefine the degree of centrality, determine new conditions of environmental continuity through the recognition of the **heritage value** available, and rewrite what exists, intervening by controlling relationships that the reuse of each of these areas could trigger on the local and **regional scales**.

6.5 Urban Agriculture

If the rapid industrial recovery after the Second World War deepened the city/country dichotomy, today the rural/agricultural pairing is no longer clearly opposed to the urban/industrial pairing. Indeed, in some cases, the idea of a happy rural/urban union arises, with forms of urban agriculture located not only near the city, producing an "**urban countryside**" [19], but also close to its borders and within it. Urban agriculture is an integrative part of the urban ecological system and can undertake an important role in the system of urban environmental management.

The experience of **urban agriculture** builds on the experience with **urban gardens**. These are created with the goal of favoring social aggregation and the constructive use of free time, recovering a direct and active relationship with the land and with nature to favor the transmission of natural cultivation techniques and knowledge. Later, they have also become a first step for the urban requalification of small areas or large metropolises. There are many Italian community administrations that have moved in this direction[14] (Fig. 6.4). In the **Caserta Territorial Coordination Plan**, the social and ecological value of peri-urban rural areas has been highlighted. Their agricultural function guarantees a high quality of open space, which, when made available to its citizens, can contribute decisively to increasing overall urban quality. In addition, rural spaces, linked to the system of green continuity in the **Naples–Caserta metropolitan area**, represent a key resource for the construction of ecological networks in the urban environment:

> *With objectives linked to the improvement of environmental quality (self-purification, microclimate regulation, permeability maintenance), biodiversity conservation, the promotion of agriculture and urban forests,*

[14] We have found significant experience with urban gardens promoted by city administrations in: Alba, Bologna, Bresso (Milan), Cinisello Balsamo (Milan), Florence, Genoa, Milan, Peschiera Borromeo (Milan), Naples, Padua, Palermo, Pesaro, Pisa, Rimini, Rome, Savona, Turin, and Saronno (Varese).

Fig. 6.4 Märkisches Viertel, Reinickendorf, Berlin. Example of urban agriculture near the Senftenberger Ring. Photo by Vittorio Vidotto

> *furnishing opportunities for recreation, environmental education, and life outdoors, the improvement or urban landscapes, ...* [20].

In reality, the urban garden has always been present in the history of Italian cities. Perhaps the historic minimum occurred in the 1960s, when the cultivation of gardens within the city began to symbolize an inferior social and economic condition, an element of landscape degradation. The rebirth of interest in garden cultivation coincides, instead, with the economic crisis that hit Europe in the 1980s. However, at the base of current amateur garden cultivation, there is not so much the need to be economical, but rather the desire to "know what we are eating." It is precisely in the last 20 years that there has been a rebirth of an old institution, the "garden without a home," i.e. gardens located within the urban fabric that do not belong to those who cultivate them, but to associations or administrations that assign them to nonprofessional growers.

A multifunctional, innovative activity grows out of the expression "**urban agriculture**," but it is rich because of its reference to a past of medieval gardens, in which ethics and aesthetics coexisted [21]. Urban agriculture therefore presents different values[15]:

1. Social (agro-citizenship), in that it provides opportunities for: intergenerational and interethnic **social aggregation** during free time; the creation of a community; partial economic support for families (food integration and sale of garden products); quality of food in terms of pleasure and health; and the possibility of protecting the territory.
2. Environmental, in that it provides possibilities for: living outdoors with a respect for nature (biological agricultural practices); experimenting with forms of environmental education (culture of recycling, respect for the environment); supporting the biological cycles of air, water, and land.
3. Cultural, in that it favors the rediscovery of biological rhythms and manual labor.

Finally, urban agriculture is suggested as a tool for **periphery requalification**, redesigning free areas and reconnecting them to **green systems** and **countryside systems**, implementing urban greenery and biopermeability, introducing new public spaces, and preventing the "*concept of collective place [from being] identified entirely with the idea of closed, controlled spaces aimed at consumption*" [22]. In this **framework**, peri-urban agriculture becomes the tool for redefining the territory, even using the force of historical memory wrapped in the traditions, uses, and values that permeate local culture [23]. New rural areas become a part of the so-called "**drosscape**" [24], an archipelago of natural spaces that are now wedged into the urban fabric of the consolidated city and urban dispersion, and often also spread out in a point-like fashion. Some authors [25] also point out that urban agriculture in the city, by helping to maintain biodiversity, could favor the development of **ecosystem services**[16] that lead to human well-being. Not only that. The beneficiaries of these services could also draw out economic advantages [26] by starting new activities and turning them into an economic advantage.

---

[15] For a closer look: Bailkey M and Nasr J (2000) From Brownfields to Greenfields: producing food in North American cities. Community Food Security News; http://it.paesaggioix.wikia.com/wiki/Orti Urbani; Urban agriculture in Vancouver: Greenskins Lab, Sviluppo sostenibile, Energia Rinnovabile e Nuova Estetica Urbana. http://www.genitronsviluppo.com/2009/08/28/fattoria-urbana-agricoltura-urbana.

[16] For more information: Cataldi M, Morri E, Scolozzi R et al (2009) Stima dei servizi ecosistemici a scala regionale come supporto a strategie di sostenibilità. In: Proceedings from the 19th S.It.E Congress, From the Alpine peaks to the marine depths 15–18 September Bolzano; Costanza R, d'Arge, R, de Groot R et al (1997) The value of the world's ecosystem services and natural capital. Nature 15, 387:253–260; Costanza R (2008) Ecosystem services: Multiple classification systems are needed. Biological Conservation 141:350–352; Dziegielewska D, Tietenberg T, ➜

6.6 Intelligent Mobility

Transportation and quality of life are closely related, and the management of mobility is considered one of the principal levers available to local administrations when imposing policies of **urban sustainability**. Infrastructure design can become a pivotal element when defining new landscape structures. From an environmental point of view, it is interwoven both with the problems of pollution and the obstruction of environmental continuity in its various forms, as well as with problems of energy [27].

Within this perspective, and following the **Aalborg Charter** in 1994, the European Union has been interested in sustainable mobility through the "Thematic Strategy on the Urban Environment" elaborated by the European Commission between 2002 and 2006. In 2007, the "**Green Book on urban mobility**" was published, along with a preparatory document containing the guidelines for the drafting of the **Sustainable Urban Transport Plans**. Within the European arena, the transportation system is always viewed in its intense relationship with the environment, in that the pressures exerted by transportation on the most fragile ecosystems are revealing and are often the main cause of environmental impact.

On the other hand, mobility services represent a powerful engine for economic and social development. Each locale presents its ideal means of fruition, which places the sense of place, the quantity of users that can be supported, and the type of transport suitable for approach and fruition in relation to one another. These relationships, if well devised, can lead to notable economic benefits for the community served, and contribute to defining the quality of life; however, if they are not well calibrated, they can become elements of functional and landscape detraction[17]. In this way, the discovery of the role that infrastructure itself can undertake triggers a virtuous interaction between the constituent matrices and materials in the territory and its producers [28].

However, the whole problem can be seen as a cost–benefit exercise, in which the main results should tend toward a correct balance between the advantages (economic, environmental, and social) and costs (positive vs. negative externality) of **mobility services**. Within the scope of allowing these challenges to be faced in a determined area, it may be useful to develop a reference framework that considers:

Seo SN (2009) Total economic value. In: Cutler J Cleveland (ed) Encyclopedia of Earth. National Council for Science and the Environment, Washington, D.C.

[17] The international research of the "Access2Mountain" project deals with these themes. For a closer look: Sargolini M (2012) Access2Mountain. Sustainable mobility in Alpine and Carpathians regions. In: Morandi F, Niccolini F, Sargolini M (eds) Parks and territory. List-Actar, Trento-Barcelona.

1. A collection of **indicators** for the impact of mobility services on the environment in a given community, with particular reference to the economic dimension.
2. A transverse reading of different pressures acting on the ecosystem of interest, often expressed in different value categories.
3. A means of recomposing the different readings so that quantifiable **assessments** can be adapted to those that are only qualitative, evaluations in which objective approaches prevail adapted to those that are more subjective.
4. A synthesis of the evaluations in a **decision support system** of open governance, allowing for the comparison of different visions.
5. A method to connect different **visions** to different **policies** in the economic, environmental, and urban-planning fields.

It is clear that all of the considerations previously mentioned and their related decisions heavily influence the organization and quality of urban spaces, considering that creating infrastructure in a territory can be, if well designed, a tool for the modernization and requalification of crossing spaces. However, the **model for mobility** in which they intersect, connecting and integrating even vastly different means of transport, must be a coherent part of an infrastructure program on the national level. In reality, in some countries, such as Italy, none of this has happened as yet. Even dedicating extensive resources to infrastructure, only a scattering of investments in "*small public works, often poorly designed or realized even worse, in the same way the proliferation of spontaneous or abusive settlements marks the territory and the city*" [29] has been seen.

In two research programs carried out on this theme (Umbria, Italy, 2009–2011; the Carpathian Foothills in Austria, Hungary, Romania, Slovakia and Slovenia, 2011–2012), profoundly different study areas were examined, in which different forms of approach, contact, and fruition were analyzed: one precisely on the speed of movement, even with international terminals, typical of logistics, trade, and finance; the other characterized by means of light fruition, with slow movement vectors, prevalently pedestrian and bicycle, ideal for looking at the landscape and nature. Naturally, the "slow" and "fast" adjectives constitute an effective linguistic synthesis to evoke limits and potential, considering that each type of movement expresses its own vision of sense and a specific testimony of human/territory belonging (Figs. 6.5–6.12).

Fig. 6.5 Maramureş, Romania. Footpath in a mountainous rural area

Fig. 6.6 Park Zempléni Tájvédelmi Körzet, Slovakia. Driveways and paths at the edge of the park

Fig. 6.7 Colfiorito, Italy. Fast-paced and isolated mountain communities on the edge of the National Park of the Sibillini Mountains

Fig. 6.8 Corvara in Badia, Italy. Mode of transport for the use of rope-tow ski basins

Fig. 6.9 Prešov, Slovakia. Fast paths crossing protected areas

Fig. 6.10 Colfiorito, Italy. New road infrastructure under construction near the National Park of the Sibillini Mountains

Fig. 6.11 B d _PSeA _PbFabGa b]cWWb[dYeb Ye dNeb db]e_Ucd Wb]a_ ea ecZde b d[SeboaeZdeU__ce_UecZd Tatra Mountains

Fig. 6.12 Melk, Austria. Fast paths that intersect the Danube river

References

1. Waldheim C (2006) The Landscape Urbanism Reader. Princeton Architectural Press, New York
2. Newman P W (1999) Sustainability and cities: extending the metabolism model. Landscape and Urban Planning, 44
3. Chiria GM, Giovagnorio I (2012) The role of the city's shape in urban sustainability. In: International Transaction Journal of Engineering, Management, &Applied Sciences &Technologies, vol. 3 n.3
4. Owens S (1986) Energy, planning and urban form. Pion, London
5. Barbieri P, Ricci R, Ulisse A (2009) Reti energetiche. In: Barbieri P (ed) Hyper Adriatica. Op2 Opere pubbliche e città adriatica. List-Actar, Trento-Barcelona
6. Doedge H P (2006) The Renewable City. Wiley Publishing, New York
7. Harvey D (1996) Justice, Nature and the Geography of differences. Blackwell Publishing
8. Camagni R (1996) Economia e pianificazione della città sostenibile. Il Mulino, Bologna
9. Bologna G (2008) Manuale della sostenibilità. Idee, concetti, nuove discipline capaci di futuro. Edizioni Ambiente, Milan
10. Bologna G (2011) Il sorpasso dei limiti alla crescita. In: Sargolini M and Talia M (eds) Prevedere anticipare, Prisma, no. 1/2011. Franco Angeli, Milan
11. Bertuglia C S, Staricco L, Rota R (2004) (eds) Pianificazione strategica e sostenibilità urbana. Concettualizzazioni e sperimentazioni in Italia. Franco Angeli, Milan
12. Di Palma V (2012) Rottamazione e rifunzionalizzazione. In: Sargolini M, Talia M (eds) Riconoscere e ri-progettare la città contemporanea. Franco Angeli, Milan
13. Di Palma V (2011) Demolizione e ricostruzione nei programmi di riqualificazione urbana. Aracne, Rome
14. Sargolini M (2002) Transferable development rights. Balancing conservation and growth. In: Project Planeco. Planning in ecological network, n 6. University of L'Aquila
15. Ferraresi G (2009) Produrre e scambiare valore territoriale. Alinea, Florence
16. Sargolini M (2009) Paesaggi agrari e continuità ambientali. Proceedings from the conference: "Verso una politica del paesaggio. Temi e buone pratiche a confronto", Umbria Region. In: Architettura del paesaggio, 21. Paysage, Milan
17. Clement G (2005) Manifesto del terzo paesaggio. Quodlibet, Macerata
18. Marini S (2011) Nuove terre. Architetture e paesaggi dello scarto. Quodlibet, Macerata
19. Donadieu P (2006) Campagne urbane. Una nuova proposta di paesaggio della città. Donzelli, Rome
20. Frisch G (2012) Tra città e campagna. In: Sargolini M, Talia M (eds) Ri-conoscere e ri-progettare la città contemporanea. Franco Angeli, Milan
21. Ingersoll R, Fucci B, Sassatelli M (2009) Agricoltura urbana. Dagli orti urbani all'agricivismo per la riqualificazione del paesaggio. Quaderni sul paesaggio, Emilia Romagna Region
22. Ciorra P (2002) Adriati-città. Un paesaggio post-industriale. In: le Cento Città, n. 21
23. Kipar A (1999) Le aree periurbane nel contesto metropolitano. Tra città e campagna. In: Camagni R (ed) La pianificazione sostenibile delle aree periurbane. Il Mulino, Bologna
24. Berger A (2007) Drosscape. Wasting Land in Urban America. Princeton Architectural Press, New York
25. Santolini R (2008) Paesaggio e sostenibilità: i servizi ecosistemici come nuova chiave di lettura della qualità del sistema d'area vasta. In: Riconquistare il Paesaggio, La Convenzione Europea del Paesaggio e la conservazione della biodiversità in Italia. Miur, WWF Italia, Rome
26. Scolozzi R, Cataldi M, Morri E, Santolini R, Zaccarelli N (2010) Il valore economico dei servizi ecosistemici in Italia dal 1990 al 2000: indicazioni per strategie di sostenibilità o vulnerabilità. Valutazione Ambientale anno IX n. 17. Edicom Edizioni, Gorizia
27. Newman PWG, Kenworthy JR (1989) Gasoline consumption and cities. A comparison of U.S. Cities with a Global Survey. Journal of the American Planning Association
28. Gasparrini C (2003) Passeggeri e viaggiatori. Meltemi, Rome
29. Clementi A, Pavia R (1998) Territori e spazi delle infrastrutture, Transeuropa, Ancona

7 Environmental Networks for Smart Territories

In this chapter, the concepts of **environmental networks** and **territories** are brought together. They are already dense with meaning individually, but we make adjectives of them and consider the added value that derives from their close geographical/**conceptual connection**. To carry out this reflection, we start from the case study of the **Marche Regional Ecological Network** (REM) for two reasons: first, considering the range of interpretations referring to this concept, by presenting a concrete case we can refer to a specific meaning, which we would like to reflect on without missing on an appropriate comparison to other approaches; the second reason is directed at the recovery of an experiential methodology that we often overlook in favor of abstractions. Unfortunately, in the field of governance, we have often seen what can sometime happen with apparently irrefutable theories, such that nothing positive is left in the territory if the comparison with concrete implementation at different levels of government is not controlled. The failures recorded in the history of **landscape planning** in countries such as Italy, which should focus its development policies on this extraordinary and strategic resource, are a warning we should not ignore: from the sharing of guidelines, programs, and master plans on the regional and national levels, it is difficult to proceed with identifying the effects of implementation on the provincial and community levels. Therefore, from the very beginning, the planning of the REM followed two different yet complementary aims:

1. Defining a complete framework for **regional ecological systems** and the relationships that govern them.
2. Providing tools to "contaminate", with appropriate structural directions, the **territorial policies** that different entities (region, province, city, and so on) implement in their own skill areas, intervening on the tools used to plan vast and communal areas.

M. Sargolini, *Urban Landscapes*,
DOI: 10.1007/978-88-470-2880-7_7, © Springer-Verlag Italia 2013

In fact, it is clear that entities institutionally dedicated to the conservation of biodiversity, not only among all protected areas, but also the environmental services of the region and province and the managing entities of the **Natura 2000 network**, cannot fully undertake their tasks if they do not find an adequate application of the tools controlling transformation activities in the territory. At the same time, amid the chaos of settlement sprawl that characterizes the **Adriatic city**, ordinary and sectoral **urban planning** (for example, the landscape plan, or everything regarding the organization of the infrastructure system, programs of rural development, special plans for waste management, mining, energy, and so on) urgently needs new directives and orientations for the design of the latent city. The definition of a clear, shared framework that is scientifically correct regarding priorities, critical aims, and project strategies thus becomes an unavoidable step in transforming the REM from a simple analysis and protection tool into a proactive one. In addition to its intrinsic value, **biodiversity** is therefore viewed under the REM for the role it carries out in the widest context of territorial management, and in particular for the contribution it makes in defining the regional identifying system, through the close interaction with the process of revision of the regional landscape plan and its adaptation to the European Landscape Convention (ELC).

In this view, comparing the environmental network to the city takes on pivotal value due to the direct and indirect functions it will be able to carry out in relation to increasing the overall quality of life of its citizens. Moreover, strengthening the combination of services provided by the ecosystems while respecting the protection of essential resources such as air, water, or land, is an element of primary importance for the development of the "green economy" recognized by the region as the master plan for facing the current serious economic crisis. In this sense, the REM can become, on the one hand, a strategic tool in the area of **development policies** put into place by different entities, configuring itself to start the process of reaching the objectives defined by the "National Strategy for Biodiversity" (Ministry of the Environment and Territory, 2010) and the "EU Biodiversity Strategy to 2020" (European Commission; COM(2011) 244); on the other hand, it can become the **basic matrix**, the framework for the formal and functional redefinition of the Adriatic city. For all intents and purposes, it is configured as the basic network in the reticular perspective within which the recent institutions of "smart cities"[1] and "**creative cities**"[2] move. For this reason, as we will see from the

[1] The debate on smart cities began ten years ago. It has often concentrated on newly forming cities, overlooking interventions on the existing city. Currently, however, it could support processes of urban regeneration.

[2] "Città creative (creative cities)" evokes the contemporary city when it is the engine of sustainable development, and therefore when it is more attractive, more dynamic, and more livable. For more information, see: Florida R (2005) Cities and the Creative Class. Routledge, London; Carta M (2012) Quel motore nelle città creative. In: Il Sole 24 Ore, 22/03/2012; Carta M (2012) Creative city 3.0 New scenarios and project. List-Actar, Trento-Barcelona.

experiments outlined in Chapter 8, the environmental network will be the first project to be implemented with the goal of providing a structural reference framework and a spatial anchoring to other transdisciplinary actions.

7.1 The Environmental Network and Other Networks in the City

In the European experience, the different environmental network projects, which are becoming ever more widespread, can be divided into two different types of approaches:

1. Those that have a vision of the environmental network bounded by a strictly ecological field, extended only to favor the movement of animals and plants, with the sole objective of finding empty spaces and passages within urbanized areas.
2. Those that consider the environmental network as an infrastructure that, in addition to guaranteeing the movement of animals and plants, brings into play all urban, peri-urban, and **ex-urban spaces** that intersect it through requalification and reorganization, favoring contact and osmosis between areas internal to the city and **natural areas** in the context, consolidated areas and areas in formation, old and new centralities, ecological continuity and social frameworks [1, 2].

The Marche REM falls decidedly within this second approach, since it has been designed to guarantee functional, cultural, recreational, and ecological connections coherently with the redefinition of the landscape of the Adriatic city. To define a register for reading the local characteristics of anthropic and **biological systems**, evaluating structure, criticality, and opportunities, and to program an appropriate design connection, the **functional ecological units** of the REM maintain a dialogue with the **landscape fields** regarding the landscape plan undergoing revision, highlighting common implementation processes (Fig. 7.1). Final design verification is therefore included in the implementation process. It is developed with 25 strategic frames divided into five themes and emerging contexts:

1. The **coastal city** and residual environmental relationships with the **hills**.
2. The settled **valley floor**, **river connectivity**, and green backbones.
3. Agrarian landscapes and the diffuse connectivity of inland areas.
4. The dilated Apennines: the transition between the ridges and the foothills.
5. The Apennine ridges and the connection between protected territories.

For each of these, the project strategy for the application of the REM has been developed, and even the definition of timely interventions has been added to the five themes. In this phase, actions have been introduced to define:

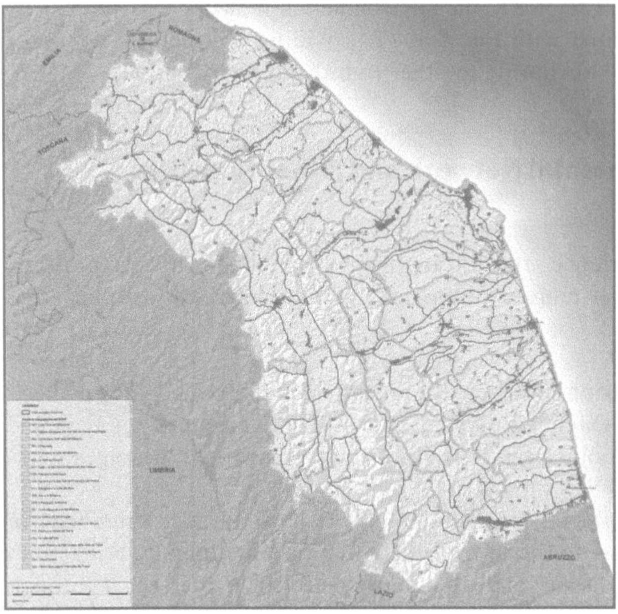

Fig. 7.1 Marche Regional Ecological Network (REM) and PPAR. Comparison of functional ecological units defined by the REM and areas of adaptation to the PPAR. *PPAR* Piano Paesistico Ambientale Regionale (regional environmental landscape plan)

- The satisfaction of several basic standards of urban quality.
- Containment lines for some industrial, productive, and **residential expansion**.
- Lines to reduce melding processes between urban contiguities.
- Indications and orientations for new expansions and infrastructure storylines.
- Ways to overcome **bioconnection interruptions**, especially close to where the environmental network intersects with various pre-existing axes, either designed or programmed (or through the introduction of wildlife crossings with underpasses and overpasses).
- The formation of tree stands, **forest systems**, and **sound-absorbing barriers** for the reduction of the impacts produced by transportation and production activities and for the redrawing of urban borders, ecotonal areas and open spaces in the city.
- The improvement of urban green areas, and sections of **riparian vegetation**, specifying the organization of appropriate forms of use related to the overall design of the city and the objectives of the network.

Consequently, in the perspective of reducing occurrences of biological fragmentation, which have multiplied in the last 20 years due to the explosion of settlement sprawl in the Adriatic city, the REM urges us to provide orientations for a new city project, while also evaluating the formal and compositional balances that open spaces, whether point-like or linear, can contribute to define, by redesigning virtuous melding among residual areas (or "white areas," as previously discussed), the greenery of neighborhoods, **peri-urban areas**, **agrarian landscapes** in the immediate inland hills and the natural basin of the Apennine ridges. In this view, it is also hoped that the insular idea of marine resources—reinforced over time due to the superimposition of the north–south railways, roads, and motorways, which have constituted a rigid interruption between coastal areas and the **inland hills** (Fig. 7.2)—is overcome.

The landscape, from an insular idea of the Adriatic Sea to its insertion in a reticular vision, presupposes appropriate management of functional, ecological, and cultural relationships with the sea. The REM therefore enters within the tangle of other city networks and the territorial context, infecting them and characterizing them. For example, important intersections or combinations are developed with the **mobility network**. As a first step, the "slow path," to which the class of environmental networks belongs completely, favors a requalification of public spaces capable of introducing new characteristics of naturalness, new spaces for socializing, and new connections in the city between zones with different **use destinations** [3]. In a more traumatic way, if not appropriately studied and planned, it tries to dialogue with the "fast path," often tangent to large conurbations and, therefore, a continuous solution between country and city. In the expression of this relationship, it should be

Fig. 7.2 San Benedetto del Tronto, Marche, Italy. The infrastructure divides coastal areas from inland areas

considered that changes in the habits and behaviors of **insiders** and **outsiders** have now caused urban spaces [4]—which are an active part of the multiple flux networks—to change, linked together by "*swaths of infrastructure connections*" [5]. Thus, environmental networks, while establishing ecological and anthropological relationships on different scales through natural and urbanized spaces with reciprocal contamination—such as interstitial areas in the built-up fabric, places of collective identification, places of local memory, still-cultivated pieces of the countryside, areas of waste or reuse, new large technical parks—still trigger contact with the **global network**. On the other hand, the relationship with **cultural networks** does not exhaust contact with the system of objects deposited in the territory, but is extended to the development of dynamic and coevolutionary relationships between the cultural and territorial heritage. The introduction of the concept of "**territorial heritage**" [6] by Alberto Magnaghi points to this direction.

7.2 Landscape and Quality of Life

It is really true that the landscape, with its constant changes [7], tends to become enriched, or grow heavier, ever more. It is becoming a type of "**omni-landscape**" [8], an "**undifferentiated and all-inclusive**" container [9]. It is not only the ELC that has opened the way to broadened horizons, but also the high expectations of resident populations, eager for an important push in the requalification of living from the landscape. Moreover, reflections on the "new urban question"[3] are strongly connected to resignifying some landscape contexts, which allows for the coexistence between natural and artificial objects, and the contextual innovation of some settlement forms as an essential lever for the city project and the qualification of its renewed livability.

Different movements and currents of thought[4] concentrate on the role that the landscape can have on improving the quality of living; attention that, especially in the **Mediterranean area**—where it perhaps first emerged—seems to have been dulled for many years. As Paolo Castelnovi reminds us, the sense of the landscape is present in our culture. It is identified in the common value given to the social and cultural relationships of the territory as they pair in their perceptual form. In this society, ever more liquid and little interested in history and common good, the landscape, from the places to the network

[3] On the new urban question, see the 4th Conference of the International Forum on Urbanism (IFoU), jointly held at Zuiderkerk (Amsterdam) and at the Delft University of Technology (TU Delft), The Netherlands, in 2009, titled "The New Urban Question– Urbanism beyond Neo-Liberalism.

[4] The "For Landscape/Landscape For (http://www.landscapefor.eu/)" website gathers observations to contribute to making social energy and business intelligence productive when directed at objectives of quality of living, organizing innovative selection, and looking to bring to light a diffuse question that already deals with quality of living.

(especially when virtual), causes geographic centralities and different ranks holding on to traditional settlement models to come to mind less so; however, the landscape can and should serve to increase quality of life by acting on social relationships, perceptions, and ecosystems. In this new perspective, the objectives of sustainable development and livability of the urban environment mutually reinforce each other.

Presently, however, while we manage to define relationships between the form of the city and systems of **functional connectivity**, between building density and bioconnectivity, between land uses and functional hybridization, we do not know how to evaluate the existing relationships between the functioning of the ecosystem, the energy and the form of the city [10, 11]. Different studies have looked into this. On the regional and metropolitan scale we bring to the reader's attention, for example, the studies by **Forman** [12]; on the urban scale those of Girardet, Nijkamp, and Perrels [13]; and on the neighborhood scale those by **Corbett** and Corbett [14], Rudin, and Falk. Despite some attempts [15, 16], there is no common conceptual framework today that allows one to compare these different approaches and consequently to move from them to the construction of a scientifically logical and politically dialogic process to make decisions of governance regarding the desirable urban form and its supporting urban politics.

What is certain is that the culture of planning—in terms of scientific statutes and disciplinary skills, professional technical capacities, administrative apparatus, legal tools and institutional frameworks, as well as cultural attitudes and sensitivity—if it is to move in this field with some hope of success, should be able to:

1. Look beyond its disciplinary walls and entrust itself to perceptual, ecological, and **aesthetic interpretations** to "*balance conflicting uses that arise on the land, in the water, and in the air in a spatial, functional, and dynamic perspective*" [17].
2. Focus attention on relationships, and social, cultural, and ethnic aspects, moving from how much the representation of the urban area evidences the existing area, to how much it suggests or evokes instead.
3. Favor the sharing of **scenarios** for the city, even in consideration of the fact that it is not always easy to objectively understand the priority of one object with respect to another, and the urgency of one project with respect to another. Sometimes the conflict is not even between public and private interests, but between contrasting public interests [18].

7.3 Greening the City

Through the landscape, the environment (insofar as it is a component of the landscape paradigm) enters into the new vision of the city. **Urban sociology** is very interested in the potential of relationships between ecological and social networks, recovering the overall biunivocal relationship that exists between ecological and **human systems** [19]. This corresponds to an essential contact that brings into play the role of nature in increasing the quality of urban life. Examinations regarding the use of greenery as an element of landscape requalification of urban and **metropolitan areas** have been developed starting from the second half of the 20th century [20]. A group of scholars from Harvard University in the USA, among them Charles Waldheim and Mohsen Mostafavi, created the first plans for large interventions on linear systems that intersect, interconnect, or limit portions of the urban city. This means **greenbelts, greenways, parkways**, and the recovery of maritime or river waterfronts.

The efforts that so many European cities, starting with London, are making to rethink and focus on the idea of greenbelts in new perspectives of the reticular city branching out into the territory, testify to the difficulty of identifying new organizational logic capable of integrating open and closed spaces, urban and rural landscapes, settlement and environmental dynamics[5]. In other cases, the creation of urban parks redefines the system of relationships between the parts. This is the case with the **Ruhr Valley** in Germany, where the recovery of existing industrial structures has favored the formation of spaces for collective activities, improving the environmental quality of the **ex-industrial area** and the entire surrounding urban area. In this new perspective, "natural" and "artificial" tend to become virtually indistinguishable. "*Nature is no longer conceived as the eternal background of human actions, but as fundamental material for the project; no longer an opportunity for contemplation, but evolutionary value that generates dialogue between project and context*" [21].

Several Italian experiences also deal with these questions[6]. The **Corona Verde project**, extended to provide an organic proposal for the territorial requalification of the Turin metropolitan area, was launched in 1997 by the Piedmont Region[7]. The focus of this proposal was the improvement of places and systems of open space, of landscape, and **ecological quality** that still characterize the settlement context, through the creation of bicycle/pedestrian connections and the precise arrangement of green areas on the urban borders. The scheme, presented at the end of 2007, proposes an integrated strategy of reorganization and requalification for the **Turin area**, which has two joint aims: ecological rebalancing (with the active conservation of natural spaces and con-

[5] In France, in 2012, with the "Grannelle II" law, the concept of "Trame verte et bleu" (TVB) was introduced in urban planning (SCot and PLU) to halt the loss of biodiversity and to re-establish the ecological continuity between natural environments and urban environments. The government of The Netherlands, in 2004, adopted the National Spatial Strategy (Nota Riumte), which connects national ecological policies to strategies for territorial development.

[6] Some provincial Territorial Coordination Plans favor the discontinuity of the local settlement system in order to ensure the functioning of the ecological network: Modena, Milan, Reggio Emilia, Bologna, Rome.

[7] A detailed presentation of the Corona Verde project can be found on the "For Landscape/Landscape For" website (http://www.landscapefor.eu) edited by Paolo Castelnovi.

nection networks, the protection of the hydrographical grid, and the defense of rural space as an essential component in the framework of life of the **metropolitan population**), and the development of the historical/cultural heritage, both in its expressions of exceptional worth and in the diffuse systems of the cultural landscape. Overturning the traditional approach, which starts from open spaces, in particular from what remains of the peri-urban countryside, it proposes new urban planning and infrastructure directions and new strategies for sustainable development.

The Marche REM also confronts these themes of contacts between the **ecological network and the city**. It concretizes the idea of a city penetrated by environmental networks that communicate with the other networks and connect green areas and white areas, contaminating consolidated fabrics and undefined city fragments, contributing, together with other infrastructure networks, to the definition of the **city structure**. Projecting environmental networks for urban areas can produce important strategies for "**green networks**" intended as "large parks" [22], within which welfare strategies unfold (Figs. 7.3–7.6).

7.4 Cities and Territories

The city is a fertile element with a vital relationship to the territory. The design of the city itself cannot make sense if it is incapable of understanding the relationships with the **territorial context**:

Fig. 7.3 Umbria, Italy. Environmental networks incorporate ancient rural villages

Fig. 7.4 Palermo, Italy. Greenery branches and penetrates into the city

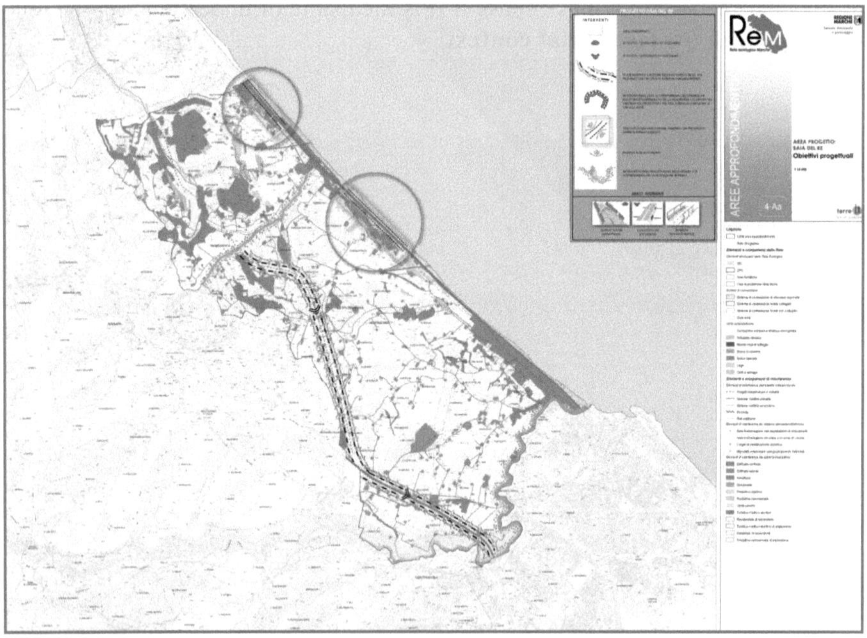

Fig. 7.5 Marche Regional Ecological Network (REM). Project Area "Baia del Re - Pesaro". The environmental network connects the urban areas of Pesaro and Fano with crags and hills creating a new framework

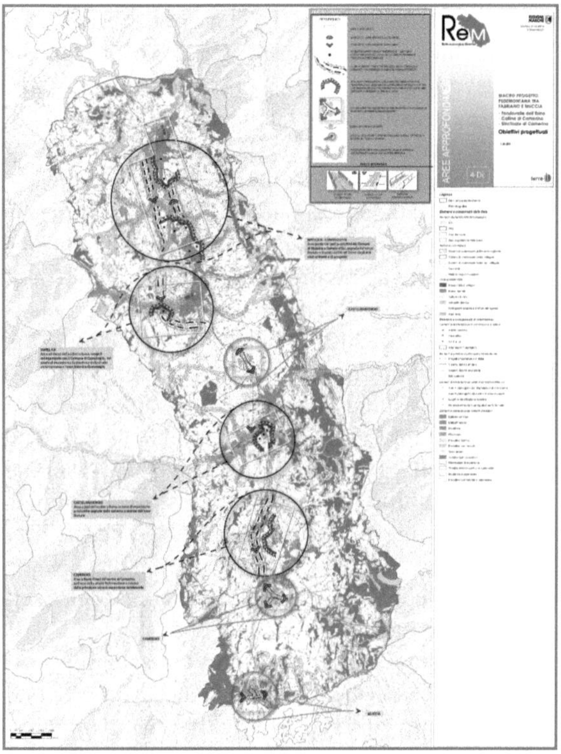

Fig. 7.6 Marche Regional Ecological Network (REM). Project Area "Syncline Fabriano - Muccia". The environmental network connects the industrial areas and residential areas of Fabriano, Matelica, Castelraimondo, Camerino and Muccia with the natural core of the foothill areas of the Appennines

> *If one traces the history of the modern European city from the 1500's renovatio urbis to the construction of the* ***industrial city*** *in the 19th century and the first half of the 20th century, one gathers how the urban project, in a strenuous effort to progressively capture the infinite, gradually expanded from the architectural object and its relationships with other objects throughout the city and territory* [23].

The territorial context is often also a "*creative resource, an openness that breaks the closure of the already defined frame and obliges us to widen our view, to integrate in the project something that was originally extraneous*" [24]. At the same time, opening the context often means making the plan/project process more complex, because one enters a space without defined **borders**. The context, coming from the outside with respect to the project frame, is not easy to understand in its deep meanings and needs. In addition, since the relationships between urban and territorial structures are confronted through the rules of the type of network, the limits they are investigated within become elusive. This does not mean a **buffer zone**; however, subject to the dominion of the **core central area**, it may be; on the contrary, it is uncontrollable in its terms.

In the **Renaissance**, man was considered to be at the center of the universe. He held himself to be ruler of the surrounding world, which he thought he could model according to his needs, and conform it to his taste. In successive centuries, possibilities of "controlling" reality have been effected through ever more subtle and sophisticated, but less concrete, methods. The **Enlightenment** thought it could maintain control through reason; **Romanticism** through emotion. The beginning of the 18th century placed man at the center of an observation system that privileged the subject; spatial centrality decreased and the self-referential centrality of the observer was favored. With the arrival of networks, the **virtual world** definitively surpassed the real world and contacts were developed despite geographical distances [25]. Territorial limitations disappeared. **Communication** became ever more widespread between basically similar groups sharing the same principles and cultivating the same interests, rather than between spatially close groups. From the geographical/territorial point of view, the effects are notable. If proximity is no longer an essential requirement to maintain social relationships [26], connections are tightened, contacts are started, and agreements are interwoven above and beyond the limits of one's own living environment[8]. The concept of neighborhood disappears, and with it, contiguous, peripheral, and contextual areas, and processes of settlement diffusion are consequently developed.

It is difficult to say if **urban dispersion**, mentioned many times in this volume, is the cause or the consequence of the phenomenon of network affirmation. The fact is that a close interweaving subsists between the two dynamics. On the one hand, the phenomenon of sprawl is identified through the urbanization of extensive rural areas close to the city; on the other, through the loss of proximity and the consequent growth of other forms of community, social, and support networks, which are often not visible at first glance. Certainly, these last items are forms of contact that lead to deep connections, well beyond the relationships between traditional centralities.

Environmental networks, while conceptually maintaining their physicality, trigger in this virtual system contacts between the city and a context that is indefinite and indefinable. The process of "scattering" the city throughout the country and the related degradation of extended portions of the **ex-urban territory**, which is losing its distinctive elements and its **rural society** relational methods, favor directing new attention at reinventing contacts between the city and the country. Environmental infrastructures in Europe are also being developed in this biocultural perspective, where contact between the city and the territory, as has already been experimented in some **North American** contexts, assumes a denser and more complex meaning that integrates nature and city,

[8] Naturally, as Bordoni C. notes in op. cit., 2009, loss of proximity is not only due to geographical causes, such as the displacement of social strata from the center to the suburbs due to the effects of changes in economic conditions, rent increases, or urban revolutions. The reasons, as always, are many, and when they all merge in the same direction, they cause tangible consequences: in this case, loss of proximity is also linked to loss of work.

Fig. 7.7 Umbria, Italy. Environmental networks of connection between town centers and shopping malls

gathering different resources and values, losing the sense of limits and object confines. Their fluidity evokes the experience of the "song routes" of the **Australian** aborigines [27]. This fluidity is not only present when leaving the city, i.e., toward territories of settlement dispersion and urban expansion, but also when entering the city, i.e., in the fabric of the city.

In the territorial **city–context** relationship, environmental networks could also be of help in finding relationships between cities and "non-places" [28] dispersed throughout the territory, often considered to be ungovernable "elsewheres" (Fig. 7.7). These autonomous, independent fragments are not different worlds separated by the generative city. If appropriately intersected and contaminated by the landscape/environmental network, they can contribute to: designing modern landscapes; re-reading territorial centralities, whose definition has been ever less so entrusted to strictly economic and functional relationships and more entrusted to symbolic relationships, identifying images, and "intangible" dynamics; again raising for discussion the idea of public space, which cannot be limited to the location of convergence and physical aggregation—the square—but is extended to open spaces, the country, and spaces of urban scattering. Therefore, environmental networks enter the landscape completely, from the society of places to the **society of fluxes** [29]. The disciplines of urban planning and architecture should study the contributions they have to offer, aware of the new mission they have before them, and give answers for governing these new identifying nodes as the sites of contact between the local communities, territorial contexts, and/or virtual and global networks.

7.5 Smart Communities

The territory, of which the city is a part, expresses complexity because it manifests a "*relationship with an active system*" [30]: **human communities** establishing their own relationships with the Earth and, moreover, determining landscapes. If we want to give life to quality landscapes, we should imagine intelligent territories, but since territories represent the space where community expectations are found, we should care for the reference communities in each territory.

Although the author does not intend to venture into the identification of reference communities, mention of the celebrated *Social Formation and Symbolic Landscape* by **Denis Cosgrove** (1984) is warranted for closer inspection by the reader. In Cosgrove's text, two ideal figures—the insider and the outsider—which, naturally, approach the territory with different value scales[9], are identified. However, the concept of community is closely linked to that of territory and therefore the attention that we direct to the territory corresponds to the well-being and quality of life of the settled community.

The desire to start from the community was expressed very clearly at the most recent **World Conservation Congress of the International Union for Conservation of Nature** (IUCN), which took place in **Jeju Island**, Republic of Korea, on 15 September 2012. Among the 19 confirmed principles of the treaty, all oriented toward a new era of safeguarding, sustainability, and solutions based on nature, there was special interest in community involvement. Some parts of the treaty are expressly directed at protecting the rights of socially vulnerable stakeholders, in particular local communities and **native peoples**, to encourage the just and fair sharing of advantages arising from the ecological functions of biological diversity; to support consciousness awareness, knowledge, good governance, and sustainable investments, to demonstrate that environmental protection is everyone's affair and that humanity is fundamentally dependent on nature; to convince everyone that the improvement of nature and ecosystems is a crucial first step in offering advantages, compensation, and recognition to nature's guardians. The IUCN will guide a conservation movement that approaches the community, **civil society**, **governments**, and investments with the goal of negotiating and implementing the practical solutions that nature offers us to face the multiple challenges of development, thus demonstrating their profitability and measuring and verifying their effects. Solutions founded on nature are based on the role of ecosystems that, if well managed, can contribute to increasing the resilience of humans and offer supplementary opportunities to all communities. This initia-

[9] For further information, we recommend: Salsa A (2009) Il tramonto delle identità tradizionali. Priuli e Verlucca, Turin; Wolf ER (1972) L'Europa e i popoli senza storia. Il Mulino, Bologna; Turner V (1976) La foresta dei simboli. Morcelliana, Brescia; Trigilia C (2005) Sviluppo locale. Un progetto per l'Italia. Laterza, Rome-Bari.

tive seems to be an effective course of action for forming smart communities, insofar as biodiversity becomes an occasion to reach broader social objectives.

The case study of the Marche REM, which is refined in the territory through contributions by different stakeholders, presupposes activating participation and communication processes that become important tools for its implementation. Tools for creating the network should, in fact, coincide with a rich characterization and articulation of those acting in its transformation. The sharing of network objectives, strategies, and measurements certainly cannot start after the planning of the REM, but is activated even from the cognitive phase. The community should contribute to interpreting the elements of crisis in the city and to "*translate individual worries into public questions*" [31]. Thus, the evaluation of the current planning and programming system is already a first step in contacting all of the stakeholders that should contribute to its implementation, as well as disseminating general objectives and specifying the real territorial effects of the REM.

The interest of the community in the network is deep, insofar as it is linked to attention for quality of life and insofar as it passes through the combination of **aesthetics** and **ecology**. For so long, a city has been imagined in which the aspiration for beauty was another thing entirely with respect to aspirations for the city's **ecological dimension**. The challenge that intelligent communities are already facing is, instead, oriented precisely at bringing together these two extraordinary dimensions. In fact, the community perceives that beauty does not derive only from **aesthetic artifice**, but also from the same need faced in project choices. It is precisely ecology, together with other territorial, economic, and **social sciences**, that argues for "the need."

The REM is therefore a basic environmental infrastructure that involves the community directly and which, placed above commonly frequented infrastructures (such as transportation or energy), tends to ensure conditions for a quality future throughout the territory, not only in terms of well-being for the inhabitants and in response to the expectations and imagination of the "**interested populations**," but also in terms of new aesthetic interpretations dictated by new ecological and symbolic horizons.

In the Adriatic city case study, which we discuss in the next chapter, the REM becomes the first project, limited to the Marche territory, shared by the local community and carried to completion. The REM has had this priority as long as the general objective has been to maintain the environmental infrastructure as the principal structure of the Adriatic city, with which particular project and governance decisions are engaged. The environmental network becomes the spatial anchor of the overall systematic vision, which is flexible and fluid, and which we address in the next chapter.

References

1. Benedict MA, McMahon ET (2006) Green Infrastructure, Linking Landscapes and Communities. The Conservation Found. Island Press, Washington D.C.

2. European Parliament (2012). EU Biodiversity Strategy to 2020. Disponibile su: http://ec.europa.eu/environment/nature/biodiversity/
3. International 139 (2009) Landscape infrastructures. Editoriale Lotus, Milan
4. Lanzani A, Marini E (1993) Il territorio che cambia. Segesta, Milan
5. Clementi A (2008) Corridoi, piattaforme, città senza fine. Nuovi scenari per la città medioadriatica. List-Actar, Trento-Barcelona
6. Magnaghi A (1998) Territorial heritage: a genetic code for sustainable development. In: INURA (ed) Possible Urban Worlds: Urban Strategies at the End of the 20th Century. Birkhäuser, Basel/ Boston/Berlin
7. Lanzani A (2003) Paesaggi italiani. Meltemi, Rome
8. Jakob M (2009) Il paesaggio. Il Mulino, Bologna
9. Gasparrini C (2012) Una nuova città: diffusione e densificazione. In: Sargolini M, Talia M (eds) Ri- conoscere e ri-progettare la città contemporanea. Franco Angeli, Milan
10. Williams K, Burton E, Jenks M (2000) Achieving sustainable urban form, E&F SPON, London and New York
11. Chiria GM, Giovagnorio I (2012) The role of the city's shape in urban sustainability. In: International Transaction Journal of Engineering, Management, & Applied Sciences & Technologies, volume 3, n. 3
12. Forman R (1997) Land mosaics: The ecology of landscapes and regions. Cambridge University Press, Cambridge
13. Perrels P A (1994) Sustainable cities in Europe. Earthscan, London
14. Corbett J, Corbett M (2000) Designing sustainable communities: Learning from village homes. Island Press, Washington DC
15. Williams K, Burton E, Jenks M (2000) Achieving sustainable urban form: Conclusions. In: Achieving sustainable urban form. E & FN Spon, London
16. Jabareen YR (2006) Sustainable Urban Forms: Their Typologies, Models, and Concepts. In Journal of Planning Education and Research, September, 2006 26: 38–52
17. Steiner F (1994) Preface. In: Steiner F (ed) Costruire il paesaggio. McGraw-Hill, Milan
18. Zagrebelsky G (2008) Valori e diritti. In: La Repubblica, 22 February 2008
19. Osti G (2010) Sociologia del territorio. Il Mulino Manuals, Bologna
20. Secchi B (2005) La città del ventesimo secolo. Laterza, Roma-Bari
21. Clementi A (2012) Paesaggi di frontiere urbane. In: Sargolini M, Talia M (eds) Ri-conoscere e ri-progettare la città contemporanea. Franco Angeli, Milan
22. Czerniak J, Hargreaves G (2007) Large parks. Princeton Architectural Press, New York
23. Secchi B (1995) Le condizioni generali. In: Casabella, 629
24. Catucci S (2012) Postfazione. In: Morandi F, Niccolini F, Sargolini M (eds) Parks and territory. List-Actar, Trento-Barcelona
25. Bordoni C (2009) Vivere in rete. L'individualismo dentro i social networks: Facebook e oltre. Prometeo, 106
26. Castells M (2002) Galassia Internet. Feltrinelli, Milan
27. Chatwin B (1995) Le vie dei canti. Adelphi, Milan
28. Augè M (2009) Nonluoghi. Introduzione a una antropologia della surmodernità. Eleuthera, Milan
29. Bauman Z (2002) Modernità liquida. Laterza, Rome-Bari
30. Gubert R (1987) Territorio. In: Demarchi F, Ellena A, Cattarinussi B (eds) Nuovo dizionario di sociologia. Paoline, Milan
31. Bauman Z (2000) La solitudine del cittadino globale. Feltrinelli, Milan

Techne. Interdisciplinary Examination for an Integrated Vision. Case Study: the Adriatic City

8

When Galileo caused balls, the weights of which he had himself previously determined, to roll down an inclined plane, and Torricelli made air support a weight that he knew to be equal to the weight of a known volume of water … a light broke upon all students of nature. They learned that reason only has insight into that which it produces after a plan of its own, and that … it must force nature to give answers to questions of reason's own determining; and must not allow itself to be kept, as it were, in nature's leading strings; because otherwise our observations, made in obedience to no previously thought-out plan, can never be made to yield a necessary law… Kant (1787) Preface. In: *Critica della ragion pura*, Laterza, Rome–Bari.

The different forms of **environmental determinism** (starting with the intuition of Ian McHarg) [1], which accompanied the interpretation of territorial dynamics in the last century, extinguished themselves without the need to refer to **Kant**'s thoughts regarding "students of nature." Finally, we are convinced that nature does not design itself, let alone areas originating from a deep complex interaction between natural and anthropic dynamics, such as agrarian landscapes or settlements. On the contrary, it is necessary to develop design interpretations capable of structuring themselves amid the **complexity of systematic visions**.

The complexity of systems that describe a territory or a city is not an intrinsic property, but depends both on the model used for the description and on the variables considered. The greater the number and variety of relationships between the elements of a system, the greater its complexity, insofar as the relationships between elements are non-linear. Today, confronting the sustainable design of the city means immersing oneself into reading the behavior of complex systems characterized as much by numerous different elements as by the numerous non-linear connections[1].

---

[1] The Santa Fe Institute is one of the most important research centers for complexity theory. It is particularly dedicated to the study of complex adaptive systems, i.e., complex systems capable of ➜

Massimo Sargolini, *Urban Landscapes*,
DOI: 10.1007/978-88-470-2880-7_8, © Springer-Verlag Italia 2013

Several different international research groups work in light of these scenarios[2]. There have been different attempts to produce synthetic visions (a synthetic framework was presented in Chapter 6.1 "Evaluating and Monitoring the Sustainability of Governing Actions"), but the different experiences shown present partial visions from limited angles. The experiments created to outline an **interpretive synthetic** approach do not seem mature yet.

With the aim of being able to make a small step forward with these experiments, research is being carried out in the Adriatic region of Italy[3]. The results expected are not closed, pre-packaged solutions, but the outlines of processes, guidelines, and essential orientations to face themes of **urban sustainability** in a structural and non-superficial way [2, 3]. With the goal of creating a decision support system of governance to increase quality of life in the city, the research activities foresee:

- Reinterpreting, using a transdisciplinary approach, the relationships between different physical and environmental, morphological, historical, and socio-economic **components of the city**. The project includes a comprehensive dynamic analysis of the urban system, with contributions from multiple **integrated disciplinary** sectors, such as ecology, geology, statistical mechanics, thermodynamics, technological sciences, economics, sociology, environmental planning, architectural composition, and urban planning.
- **"Measuring" the sustainability** of the urban environment by studying a dynamic model that updates and surpasses traditional indicators based on the vertical nature of the themes under investigation: mobility, building construction, **energy efficiency**, air and water quality, environmental quality, and so on. Acting separately, these have been shown to be generally insufficient at understanding urban complexity and ineffective in proposing constant and cyclic elements of support with respect to strategic/structural and operational planning tools. A holistic and **synthetic vision** also requires the introduction of a parameterization system capable of consider-

adapting and changing following an experience, for example, living organisms characterized by their evolutionary capacity: cells, organisms, animals, humans, organizations, societies, politics, and culture (For further information: Holland JH (2000), Emergence: From Chaos to Order. Oxford University Press; Gell-Mann M (1996) Il quark ed il giaguaro. Bollati Boringhieri, Turin).

[2] As an example, the author mentioned sustainable sensitive urban design, which, in different experiments, redefines the project contents in relation to the complexity of intertwining among different levels of sustainability.

[3] The transdisciplinary research group involves colleges from the University of Camerino and other Italian universities (School of Advanced Studies, University of Pisa; University and Polytechnic of Turin; University of Urbino) and foreign universities (University of Minho, Portugal; University of Gjirokastër, Albania; University at Albany–SUNY, USA). The core from which the first ideas for the development of the research program arose is composed of: C. Bisci, G. Cantalamessa, R. C. Grifoni, R. D'Onofrio, F. Ottone, M. Perriccioli, M. Talia and M. Sargolini.

ing the variability (within a predefined range) of quality-of-life indicators to envisage a formalizing process for relationships and constraints that is capable of analyzing such indicators to determine their connections.

- Applying the new evaluation and decision support model to both local- and **wide-scale plans**, with the aim of being able to evaluate the development scenarios contained in the plans, and relate them to the objectives of sustainability that each territorial area plans to reach.

The use of **parameterization systems** suitable for the construction of a decision support model that uses the indicators selected in an aggregate form is probably an innovative step with respect to the experiences observed up to now. In mathematics, a **paramete**r is an arbitrary constant used in systems, formulae, and equations (called, appropriately, parametric) that can assume any value for a determined field (set of numbers). The introduction of **parameterization** could give the system the flexibility it needs in the specific area of application. The number of parameters necessary for the description of a system is directly proportional to the system's complexity [4, 5].

8.1 An Operational Decision Support Tool

To define the model, the "Adriatic city," stretching from **Veneto** to **Apulia**, was defined as the study area. It is an area characterized both by settlement sprawl that moves from the coast toward the inland areas without visible continuity solutions, and by strongly negative elements: poor air quality, building and open spaces, the lack of easily usable **green areas**, mobility in a state of permanent congestion, elevated levels of noise pollution, the lack of adequate production management, and the lack of waste collection and disposal management.

The model considers the existing connections and relationships between the different areas of investigation identified to deal with the problems mentioned previously. These are areas that are generally represented by the usual indicators of environmental sustainability. The objective was to identify their mutual correlation and adaptability to the context of the Adriatic city (process of parameterization). From among the various parameters, the model should identify those that best interpret strategies for decision-making, governance, and the policies for change indicated as an objective. Once the synthetic measurements through a parameterization model that will regroup and synthesize the set of preselected indicators are obtained, these indicators become the basis for the development of a decision support system to be applied in the course of the policy cycle and throughout its different subsequent phases to reach a set of predefined objectives: landscape quality, environmental comfort, and energy efficiency for a better quality of life [6, 7]. With such an aim, we proceeded through the following steps:

1. Identify those sustainability indicators and procedures, known and already tested by each research unit and with reference to multiple urban sustainability paradigms (land consumption, **environmental comfort**, **energy efficiency**, quality of the natural environment, landscape, building quality, and so on), that manage to describe better than other indicators the territory of the Adriatic city in its specific characteristics and components, and which best compare with other known indicators.
2. Construct a system for collegial comparison of the indicators identified in the first phase with the goals of:
   a. Selecting those that are actually useful (because they are sufficiently integrated and easily applicable) in quantifying the components of the model we wish to construct.
   b. Constructing or adopting the indicators where (lacking consolidated references in the literature) they are seen as useful and convenient for signaling and measuring the interrelations between the effects of different phenomena on the territory, which are at times antithetic.
3. Create and test the model, with the scope of evaluating the interpretational and/or predictive capacities. Testing in this phase is performed on some "sample territories" in the Adriatic city, with contributions from stakeholders and interested deciders, to furnish them with usable information that can be integrated with real policies. In particular, the model is applied to a scenario "0" coinciding with the current state of the territory under examination, and to a scenario "1" coinciding with the provisions contained in the local or large-scale plans under consideration.
4. Evaluate the imbalances in the current state and the effects on quality of life from the urban transformations programmed by the planning tools under examination. Research—via planning simulations built with the support of the **model**—an **equilibrium condition** between scenario actions foreseen or foreseeable for the future and the objectives of sustainability and quality of life assumed by the territory.
5. Refine the model following the first experimental application, with the aim of calibrating and optimizing its performance with respect to the selected indicators and the results expected.

The model becomes a decision support system that is easily usable by public administrations and interest holders in the Adriatic city and can be replicated anywhere with the right modifications. It allows different **development scenarios**, policy proposals, or present or future plans and projects to be compared with the aim of establishing the most satisfactory intervention hypothesis to be implemented, providing for all possible improvements in the successive monitoring phase [8]. In support of this tool, a **protocol** is made available to establish themes of interest, **general strategies**, and the need to promote the sustainable development of different cities in the Adriatic area, which should be shared and made operational by the **entities** and **institutions** involved, and which will find a useful verification and evaluation tool in the application of the model. The results expected from the construction and sharing of the model, and from its applicability within plans and projects, are the following:

- Ensuring greater sustainability and sharing of the changes planned and those to be planned through a transdisciplinary reading of the Adriatic territory.
- Favoring inclusivity and the implementation of policies and interventions of territorial worth, with the aim of guaranteeing the economic, environmental, and social sustainability of transformation interventions aimed at the development of local communities.
- Making the **cooperation** between stakeholders and actors that govern the territory more active, so that they can use the model as a tool to monitor complex environmental systems.
- Improving the quality of the landscape, and with it the quality of life of the inhabitants, by intervening on three large domains: (1) "distinctive and pleasant;" (2) "efficient and nice;" and (3) "clean and healthy."

8.2 Overcoming Disciplinary Walls

The principal challenge today is to understand if cities can become places to experiment with new, more sustainable forms of living and working through the regeneration of the urban landscape, which is intended, in adherence to the European Landscape Convention (ELC), as a visible, communicable image of the complexity of relationships between human and natural dynamics.

To have some hope of success in this challenge, it is necessary to follow the route of transdisciplinarity. Urban planning should break the vertical nature of knowledge and make an effort to look beyond its own disciplinary walls. This is particularly necessary in the perspective of inaugurating a new season of "land projects"—already outlined in the 1980s by **Bernardo Secchi** and proposed again 20 years later [9, 10]—in which the multiplicity of open spaces and their potential for relation, concatenation, and fruition is put into play [11].

Unfortunately, there are still many representatives from architectural and urban design that believe they can define urban transformations exclusively with their own knowledge and skills. It would be useful to ask why, in the international field, large design studios are notably **interdisciplinary** (i.e., composed of ecologists, botanists, geologists, agronomists, chemists, sociologists, economists, anthropologists, in addition to architects/land planners, who work daily side-by-side and are not simply called if necessary), while in some European countries, Italy among them (whose identifying characteristics are determined by an interweaving of different substances, whose interpretation should, naturally, occur through different bodies of knowledge), all of this still does not happen.

One sometimes objects that this state of building degradation is quite pronounced in European countries where little space is left to the **architects**; i.e., where professional figures of medium-to-low profile are approved to produce transformative projects, even those of great importance [12]. The Italian problem is rather of a different kind and resides precisely within our architectural departments. **Architecture students** can reach the end of their academic careers and graduate with perfect marks without even having the sense of the effects that their urban composition designs may have on the environment undergoing an intervention. The same simplifications of the recognition phases, which we sometimes introduce in **laboratories** to dedicate more time to design experimentation, should not lose track of the irreducible complexity of the systemic vision of the territory. This aims to put students on guard with respect to the unavoidable, deep interactions that exist between a building intervention (which is never banal and to be underestimated) and the general balance of the environmental and landscape context in which the intervention occurs. "*Because I am the size of what I see and not the size of my own height*," **Fernando Pessoa** said more than 30 years ago; and still we continue to stop each reflection at our "height" [13].

8.3 Areas of Examination

To favor overcoming the sectorialization of fields of action in this research, three large areas for further investigation have been delineated, where the different disciplinary sectors have developed integrated interpretations.

The section titled "**Distinctive and Pleasant**" closely examines the theme of open and green spaces (linear and point-like) in the city and the immediate context in their relationships with the environmental infrastructure of the surrounding territory, architectural and archaeological resources, centrality, and places of the city's collective identity. It aims to improve the activities of living, and therefore work, home, encounters, social relations, and leisure. In particular, relationships among the following are analyzed: representativeness of places in the diffuse city and in the collective imagination, giving particular attention to cognitive and **descriptive activities** (Vincenzo Riso); the common sense of places and the subjective and/or collective consciousness raising of local populations (Paolo Castelnovi); the role of natural and urban parks for **public enjoyment** and the development of new forms of fruition of **open spaces** (Roberto Gambino); cultural heritage and the environmental, social, and economic fabric in which it is located (Maria Teresa Idone); **environmental networks**, urban agriculture, and the regeneration of degraded urban and peri-urban environments (Ilenia Pierantoni); slow paths and urban networks on different urban and territorial project scales (Roberta Caprodossi and Paolo Santarelli); green urban areas and forest continuity (Roberto Canullo, Giandiego Campetella, Stefano Chelli and Marco Cervellini).

The "**Efficient and Nice**" section of this chapter examines themes of the city's overall organization and its energy needs on different building scales and in relation to urban and **economic planning** in the city and its territorial context. It aims to improve the quality of building works and related open spaces as a function of an overall reduction in energy consumption and a broader and more intelligent use of energy from **renewable sources**.

In particular, relationships between the following are analyzed: settlement dispersion and urban economies (Massimo Battaglia and Marco Frey); **sustainable mobility** and the urban economy (Luca Cetara); intelligent mobility systems and the reduction of traffic congestion (Renato De Leone); ecosystem services and the use of urban lands (Riccardo Santolini); the form of the city, urban sustainability, and the quality of the landscape (Rosalba D'Onofrio); the form of structures, technologies for sustainability, and **environmental quality** (Federica Ottone); renewable energy and the **thermal comfort** of open spaces (Roberta Cocci Grifoni).

The "**Clean and Healthy**" section of this chapter examines the themes of safety and the quality of the main components of the urban environment (air, water, land) on different organizational scales in the city and in relation to the territorial context, the requirements for living, formal balances, and both the overall and individual realizations with the aim of favoring the reduction of pollutants of any kind and increasing the quality of the urban environment. In particular, the relationships between the following are analyzed: waste and environmental health (Barbara Fenni and Sara Spuntarelli); climate variations, morphometric characteristics and urban coastal dynamics (Carlo Bisci); the morphology of places and the dynamics of the mountainside/river/coastal system (Bernardino Gentili); coastal dynamics and coastal naturalness (Alessio Acciarri, Carlo Bisci, Gino Cantalamessa and Giorgio Di Pancrazio); indicators for air and water quality and the well-being of inhabitants (Beatrice Marinelli); evaluation methods for environmental quality and quality of life (Piersebastiano Ferranti).

References

1. Mc Harg J (1969 - 1st ed) Design with nature, Natural History Press, New York
2. Bosio M, Frate M (2010) Strategie per il progetto della città sostenibile. Marsilio, Venezia
3. Owens S, Cowell R (2011) Land and limits: interpreting sustainability in the planning process. Routledge, London, New York
4. Xing Y, Dangerfield B C (2011) Modelling the Sustainability of Mass Tourism in Island Tourist Economies. In: Journal of the Operational Research Society, 62(9): 1742-1752 http://www.seek.salford.ac.uk/user/profile/publications/view.do?publicationNum=26271
5. Dobranskyte Niskota A, Perujo A, Pregl M (2007) Indicators to Assess Sustainability of Transport Activities. JRC Scientific and Technical Report - European Commission Joint Research Centre Institute for Environment and Sustainability, 2007 - 59
6. Thanh Van Mai and Kambiz E Maani (2010) Systems Thinking for Sustainable Tourism in the Cat Ba Biosphere Reserve of Vietnam. Proceeding of the 28th International Conference of the System Dynamics Society. Seoul, Korea, July 25 – 29, 2010 http://www.systemdynamics.org/conferences/2010/proceed/papers/P1312.pdf

7. Thanh Van Mai and Kambiz E Maani (2011) Toward a shared systems model of stakeholders in tourism development in the Cat Ba biosphere reserve of Vietnam. Proceeding of the 29th International Conference of the System Dynamics Society. Washington DC, July 25 – 29, 2011 http://www.systemdynamics.org/conferences/2011/proceed/papers/P1201.pdf
8. Mesev V (2003) Remote sensed cities. Taylor & Francis, London
9. Secchi B (1986) Progetto di suolo. In: Casabella, 520-521/1986
10. Secchi B (2006) Progetto di suolo 2. In: Aymonino A, Mosco V P (eds) (2006) Spazi pubblici contemporanei. Architettura a volume zero. Skira, Milan
11. Gasparrini C (2012) Una nuova città: diffusione e densificazione. In: Sargolini M, Talia M (eds) Ri-conoscere e ri-progettare la città contemporanea. Franco Angeli, Milan
12. Erbani F (2012) Alvaro Siza. In: La Repubblica, 24 August 2012
13. Pessoa F (1986) Il libro dell'inquietitudine di Bernardo Soares. Feltrinelli, Milan

8.4 Distinctive and Pleasant

8.4.1 Landscape Design Practices as Representation and Reclaiming Territory Experiences

Vincenzo Riso

The phenomenon of endless urbanization has, almost everywhere in Europe, deeply altered the original subtle territorial fabric and turned it into an ever-ending succession of single constructions, where unbuilt spaces become just meaningless rests. This is even worse with regard to the resulting landscape observation, when considering the absolute autonomy and refuse of dialogue that each single built object demonstrates.

Faced with this situation, current field experiences in the Spanish towns of Yecla, Badalona, Girona, and Costantina (Seville) [1] have shown how the valorization of primary neighborhood activities could be considered as a useful tool to redeem the typical fragmentation characterizing much of the so-called **diffused urbanization**.

Bearing in mind the principle resulting from those and similar experiences, this chapter aims to present what could be, in terms of a coherent territorial design exercise, the role of representation practices toward the implementation of a material continuity in the perception of the territory, i.e., how to arrive to the organization of what has been defined a "**landscape infrastructure**" [2].

Generally speaking, it could be stated that, within contemporary architectural design practice, the Porto School of Architecture in Portugal has, since the 1980s, maintained and promoted classical drawing as the preferred instrumental practice, to recognize the formation and transformation processes of a site and at the same time the privileged route to finding the solution. Recently, in the nearby School of Architecture of Minho University, while adopting the same drawing practice, we have also tried to extend its application to the **territorial scale**.

In principle the problem is that, when looking at the territorial scale with an architectural approach, the reality is not entirely visible or, i.e., it cannot be included within a comprehensive view. Therefore, skillful drawing observation needs to be integrated with **technical knowledge** of the specific territorial comprehension. "*To design and select, select and interpret and propose*" [3]; relying on this principle, originally defined by M. Sola Morales at the School of Architecture in Barcelona, we try to reach the elaboration of a somehow calligraphic cartography, aimed at recognizing and describing the formal and logical values of a determined territory.

Therefore, the main criterion is that of verisimilitude, i.e., the operative imperative is to maintain, even in territorial technical drawing, a kind of fidelity between reality and its correspondent representative sign, i.e., a traditional drawing standard.

Thus, we have experienced how a critical representation of a certain territory could be developed by the elaboration of a specific interpretation of the existing structures and systems, through the initial assumption of a key idea.

Incidentally, the choice of such an idea to establish the critical path for the **drawing representation process** might also be helped by the intuitive photographing of selected emerging elements, during the fundamental phase of **site exploration**, which somehow corresponds to the search of a specific—though partial—narrative of interpretation.

In the constant analysis of land use practices and forms of a given territory, the **landscape design process** (aimed at figuring out the solutions to the previously mentioned problems) is then proposed as the answer to specific site problems, but without losing the consciousness of the complexity of the global view. Then, the designation of specific **planning procedures** corresponds to the organization of an execution program and the evaluation of its different phases. Of course, devising the intervention strategy is more important than the form of its implementation, and it is within each strategy that the points and the dimensions of interventions are recognized.

Realization of the intervention strategy must be founded on uninterrupted thinking about the site. In doing so, any strategy results will be imprinted by the understanding of the initial inspection. Therefore, the spatial organization of a place can be assumed in the continuity of the meanings that it represents, but, at the same time, it is also possible to deal with the indeterminateness of the program.

An example of this, resulting from our work in the valleys of the Northern Portugal regions of Ave, Cávado, and Minho, is the recovery of the old water lines agricultural systems—and all related structures—which have emerged as real possibilities for working on the relationships among the built elements and to reclaim a sense of continuity of the land.

It can be argued that critical representation lies at the heart of understanding and at the same time imagining potential for reclaiming territory, that is an emerging landscape design practice in many other parts of the European context; moreover, in a time when, instead of creating new constructions and infrastructures, increasing effort will have to be directed toward recycling or upcycling what has already been built.

References

1. Morales Yago FJ (2007) Agua y paysaje en Yecla. Ayuntamiento de Yecla, Yecla
2. Beigel F, Christou P (1997) Paisajes épicos. In Bru E, Adell G, Schneider J, et. Al. Nuevos Territorios. Nuevos Paisajes. MACBA/Actar, Barcelona, pp 188-201
3. Solá Morales M (1979) La cultura della descrizione/The culture of description. In Lotus International 23. Electa, Milano, p 32

8.4.2 Sense of Place and Collective Identification

Paolo Castelnovi

Until its definition by the European Landscape Convention (ELC) ("*a landscape means an area, as perceived by people …* "), "landscape" was expressed as a basic contradiction. It was assumed that the perception and the consequent sense of the **landscape**, deriving naturally from a completely personal and subjective process, was instead a collective attribute of the entire population.

This is an ambiguity that runs through all political strategies underlying the ELC: in it, the landscape is proposed as a "*foundational factor in the identity of populations*," i.e., a common good that develops a **sense of community** and local differences. But we know that this common good rests on a thousand interpretations generated by subjective perceptions. These interpretations are naturally homogenized in the **society of a country**, whose members live in places that change slowly over a lifetime, but they are highly differentiated where the society or the places have recently been reorganized or have experienced upheaval. Youth, natives, or immigrants in a metropolitan hinterland have a very different sense of landscape[4].

Metropolitan hinterlands are territories where the melting pot of the population or the standardization of built spaces is found, and where the loss of local differences does not establish a deep-seated relationship between the places and those who frequent them. Cultural differences and the banality of new places make the landscape regress away from living matter within our collective dwelling, on which we found values and conduct common battles with purely ideological aims and which returns to old meanings in the nostalgic framework of an idealized relationship with **nature** [1], which is solipsistic and impotent. It sterilizes the vital potential of the relationship between people and territory, removing an essential resource from the strategy at which the ELC should be directed, precisely to give hope to those banal places and to uprooted generations.

Therefore, a "common" sense of the landscape would be a resource for the value of **identity** of our living, but only if it derives from political and planning works. One maintains that where the landscape is little integrated with the living community (or where the living community has disintegrated), processes should be triggered to render collective a sense that would otherwise remain entirely subjective, and generate radically different and contrasting outcomes [2, 3]. It is believed that involvement in a common enterprise activates a process that changes our personal sense of identity as

[4] See: De Nardi A (2010) Il paesaggio nella costruzione dell'identità e del senso di appartenenza al luogo: indagini e confronti tra adolescenti italiani e di origine straniera. Doctoral thesis, Padova. Many other contributions reflect on this theme in the online collection by Castiglioni B, De Marchi M (eds) (2009) Di chi è il paesaggio.

a result of working together with others. And if the common enterprise deals with relationships with the territory we inhabit, the product is a piece of our **"common" sense of landscape**, which acquires value as part of our personal and collective identity. Only then, will we feel proud to inhabit the place: the landscape will become part of our personal cultural heritage.

But the "common attribute," beyond qualifying a collective enterprise (ours or our ancestors'), is also an indicator of a nondedicated, nonscientific consideration connected to a diffuse sense and to the ordinary, daily (dis)attention that is normally reserved for the context. The common sense of the landscape becomes an ordinary feeling with time; when the enthusiasm ends, the memory of the commitment blurs and remains only a vague sense of belonging that is reawakened only by the "not-in-my-backyard syndrome," when they come to tell us they are going to traumatically change our world. The ordinary sense of the landscape, without a plan or attention, does not unleash infinite ordinary destruction, the erosion of recognizability and memorable signs that can be registered on the territory every year but which we do not register.

In reality, a symptom of unease can be read in the exponential growth of **tourism**, which is not explained simply by a possible increase in spending: also, landowning inhabitants 100 years ago only rarely moved from their beautiful houses and cities, which they enhanced to make pleasant.

Instead, tourist fluxes in the "cities of art" can be explained by a need for **public spaces**, signs of a **community**, and places that have such a unique identity that not only the inhabitants but also visitors take part in enjoying it. With the defeat of modernity, the cycle of identity between personal and collective is broken down: metropolitan inhabitants, free of a collective identifying recognition of the places where they live, try to reconstruct a **landscape identity**, feeling themselves to be citizens of the world, not looking for a homeland, but rather a diffuse heritage, a collection of places to which they can entrust their personal identity.

In visiting other people's places, there is a sense of respect for and recognition of the dignity of others in living well, and happily, that can sometimes even turn into envy: others are superior or at least equal to us as inhabitants of the Earth because they own their landscapes.

Perhaps this is just the latest sign of the prevalence of agricultural humans over hunters, but it is nice to think that the first Greek written text, founding our civilization, is not an epic text, but rather *Works and Days*, which Hesiod committed to paper to testify the importance of being in places and managing our relationship with nature. Every Odysseus since then refers to this when he prefers "rocky Ithaca," marked by the bed and throne made with his own hands, to any other place where they tried to detain him as an inert luxury guest without a purpose.

References

1. Raffestin C (2005) Dalla nostalgia del territorio al desiderio di paesaggio. Alinea, Florence
2. Dematteis G (2000) Il senso comune del paesaggio come risorsa progettuale. In: Il senso del paesaggio. IRES, Turin
3. Castelnovi P (2004) Il paesaggio limite del piano e del progetto. In: Quaderni della Rivista on doctoral research in landscape planning, Florence

8.4.3 Parks and Public Enjoyment

Roberto Gambino

In 1857, F. L. Olmsted signed the project for the world's most famous urban park, Central Park in New York, in the heart of Manhattan. A few years later, he worked to promote and guide the establishment of USA's first large natural park, **Yosemite National Park** in California. This noncasual concurrence reflects and symbolizes, in the work and thoughts of one of the fathers of environmentalism, the role of parks in the construction of the modern industrial city. The polarization between urban parks and natural parks characterizes the birth of the most effective tool for celebrating nature in modern territorial development.

What combines and makes the two **territorial figures** complementary is their "double mission:" the **conservation of natural heritage** and **public enjoyment**. In the US lesson, as in subsequent European experience and as an important part the rhetoric that accompanied **park policies** in the last century, that double mission was full of meaning. On the one hand, it meant that conservation was directed not only at the preservation of specific resources, but also at the celebration of natural landscapes, their value, and their symbolic and representative function, often with clear "monumental" worth, analogous to the value of cultural heritage. Additionally, it guaranteed public enjoyment of "common goods," saved from individual exploitation and directed not only at the promotion of physical **contact with nature**, in view of the well-being, health, and rights of citizens, but also at their spiritual elevation and civic education.

The classical mission of **parks** has been confirmed, not without deficiencies, conflicts, and contradictions, in the evolution of national and international legislative and institutional frameworks for the protection of nature, and, more generally, in new environmental and citizenship rights. That mission has been put to the test due to the explosion of the environmental question in all of its forms, and to the processes of globalization that drive economic, social, and cultural dynamics. The conservation option has clashed with processes of degradation and ecosystem fragmentation that threaten biodiversity at all levels and with the retreat of universal values in contemporary society. The protection of public enjoyment has become much more difficult in relation to the diffusion and aggressive growth of different forms of exploitation of the land and natural resources, including developments in tourism and energy production, which were intended and are intended to be pursued.

The difficulties are also aggravated by the spectacular growth in the number and extent of parks and other "protected areas," mostly registered in the second half of the last century, which has led to protected surfaces now covering an important fraction of the land (19% in Europe). It is a growth that has crossed ever more often (above all in less developed countries) instances of conservation with economic, social, and equity problems, and with the developmental rights and needs of vast populations. For territorial governing institutions, this carries with it the need for very different responses to emerging questions, whether within or outside the protected areas. The **International Union for Conservation of Nature** (IUCN) has moved in this direction with a proposal for the classification of protected areas within six categories, each of which contains specific planning and management strategies, above all regarding public enjoyment. Other indications have been offered by the IUCN's "new paradigms," launched at the 2003 World Park Congress held in Durban, with the goals of profound innovation in the objectives and forms of management, and the increased involvement of local populations in regulating the use and fruition of the protected resources.

The most significant aspect of the emerging orientation in park policies deals with the landscape, both within parks and protected areas—in particular, those that can be assigned to the "**protected landscape**" category, which in Europe covers more than half of the protected surface—and outside, in territories connected by important ecological, cultural, economic, and social relationships. The adoption of "landscape scale" plans in the framework of approaches to ecosystems extending largely outside their perimeters, is one of the ways through which we can try to confront the "**insularization**" of the protected areas, the fragmentation of territories, and the loss of "**ecosystem services**" that are useful, and in some way indispensable, for the sustainability of human settlements. In this direction, the landscape paradigm proposed by the Council of Europe through the ELC in 2000 has found growing confirmation in theoretical reflections and applied experiences. This highlights some key aspects, such as the recognition that the entire territory, and not just landscapes of exceptional worth, merits appropriate treatment, that the landscape value also concerns the **identity of the places**, and that landscape policies cannot in any way disregard the perceptions, expectations, and projects of local communities. Building a bridge between nature and culture requires the identifying grounding of environmental values and the overall reconsideration of the quality of living contexts, heavy with ecological, historical, economic, and cultural awareness.

The new paradigms induce to cultivate a perspective vision that faces, on the one hand, the confusing and imposing emergence of new innovative forms of "urban centrality," and on the other, the undeniable need to "bring nature back to the city," with the construction of new landscape and environmental networks branching throughout the entire territory.

8.4.4 Cultural Networks and Heritage in the Environmental Framework

Maria Teresa Idone

In recent years, the concept of heritage has increasingly caught the eye of the scientific and public communities. The traditional Western idea of **heritage** has expanded, consequently increasing the complexity of the various categories attributed to this concept as well as the evolution of those actions aimed at its preservation and at increasing its worth. This change is in part caused by the progressive convergence of the heritage with the territory and its environmental, social, and economic fabric. The terminological shift in focus from the object to the context—to the point that one may now speak of ***territorial heritage***—is also attributed to the notion of landscape established by the ELC (2000); this marks a turning point that emphasizes the need to move past single objects so as to better understand the dynamic and coevolutionary relationships that exist with and within the territory.

The territory can be seen as a "**network of networks**." Applying the network paradigm to the cultural sphere has led to the creation of policies focused on actions designed to increase the value of a given territory. In this way, cultural resources have been inserted into the economic, touristic, and cultural exchange circuits.

Heritage is therefore recognized as having an active role in the process of sustainable development. At the European level, new tools have been developed, including the **Programme of Cultural Routes** (2007), which has become an essential tool to raise awareness of the shared European heritage, as well as a means to improve the quality of life and as a source of social, economic and cultural development. International directives (e.g., **The Paris Declaration**, 2011) also highlight the close relationship between cultural heritage and sustainable development and promote a development process that incorporates tangible and intangible cultural heritage as a vital aspect of sustainability.

It is yet to be understood how the contribution of cultural heritage in attaining sustainable development effectively manifests itself, or rather how the actions of cultural heritage are tied to planning processes used to guide the transformations toward sustainability. The sphere of cultural heritage thus becomes an interesting area within which to experiment, and landscape ecology offers its contribution to the advancement of this aim by acting on the notion of cultural heritage and the connection between cultural heritage and planning.

The first is developed by shifting attention from the object to the act that the object transmits, in which its relationships with the past, the territory, and society resides. In this way, the idea of "heritage as a process" is configured as a cultural practice that constructs cultural values and meanings [1]. The second is developed by using the "landscape services concept," which used to link landscape ecology knowledge to the field of collaborative landscape planning. The services concept expands the current paradigm of landscape ecology and, in the structure–function–value chain, landscape functions can also be considered from the perspective of the values that humans attribute to landscapes [2]. Within this context, the cultural heritage, insofar as it is a constituent part of the **landscape**, offers a service that is necessarily connected to society's aspirations, and it translates them into acts aimed at satisfying this service. Therefore, heritage no longer regards only the past or a memory of the past, but it also regards the present. What we expect from it and how we achieve what we expect are planning questions that are related to the needs of contemporary societies [3].

Therefore, it is possible to conclude that **cultural heritage** is connected to the improvement of quality of life in two ways. The first because it is actually a part of the landscape itself, in which not only environmental factors contribute to the improvement of the **quality of life**, but cognitive and imaginative factors make their contribution as well [4]. In fact, one may speak of biodiversity but also bio-cultural factors (on which the IUCN also focuses) and those of "chrono-diversity" [5], which refers to temporal layers and material traces on which historical disciplines focus. Moreover, the landscape and its historical component both contribute to keeping the connections between society and their places alive and active. They go on to generate responsible attitudes and behaviors that are indispensable for the attainment of the joint objective of sustainability and development.

References

1. Smith L (2006) Uses of heritage. Routledge, London
2. Opdam P, Termorshuizen J (2009) Landscape services as a bridge between landscape ecology and sustainable development. Landscape Ecol 24:1037-1052. doi: 10.1007/s10980-008-9314-8
3. Ashworth GJ (2008) In search of the place identity dividend. Using heritage landscapes to create place identity. In: Eyles J, Williams A (eds) Sense of Place, Health and quality of Life. Ashgate, Burlington, pp 185-196
4. Sargolini M (2012) Temi ricorrenti. In: Sargolini M, Talia M (eds) Ri-conoscere e ri-progettare la città contemporanea. Franco Angeli, Milan
5. Azzena G (2011) History for places. In: Maciocco G, Sanna G, Serrelli S (eds) The Urban Potential of External Territories. Franco Angeli, Milano, pp 196-227

8.4.5 Environmental Networks, Peri-urban Agricultural Areas and Communities

Ilenia Pierantoni

The traditional relationship between city and country is one of the first issues to be addressed and redefined. Historically, this relationship used to be characterized by balance and original relationships and specifications between the city and its surrounding environment; however, it has undergone a gradual process of change and disequilibrium: the urban dimension has increasingly become the predominant element within a context of open spaces that is gradually losing value and meaning [1].

The current urban debate, after years of analysis of the nature and model of growth of the contemporary city, is finally focusing its attention on the research for new orders, rules, and alternative models capable to restore a new order in territories of extreme complexity, characterized by the constant presence of naturality, rurality, and urbanity.

In this sense the conceptual overturning and the search for a new order should start from the **environmental networks**, in particular from agricultural practices and their relationship with urban areas. Environmental networks, such as greenways [2], may significantly contribute to the regeneration of **degraded contexts**, stimulating the upgrading of natural ecosystems and the maintenance of **ecosystem services** [3]. They can specifically guide the redefinition of urban design, going into the planning and defining the structural invariants on which the new plans will lay the foundations for innovative and virtuous models of development, exploiting and upgrading territorial values. Open spaces, through agricultural innovation, may develop environmental performances and find new ways to provide public goods and spaces, with the aim to improve the quality of life; at the same time, urban areas may constitute a relevant experimental opportunity to stimulate innovative practices, such as multifunctional agriculture, to be exploited in relation to the supply of public services and products for the city. This, in a time of crisis like the present one, could lead to the development of an economic system based on the local scale, and respond to the growing demand for new social and natural spaces and landscapes, while also increasing the quality of life of **local communities**.

Therefore, **agriculture** can effectively be the regenerating element of the complexity of peri-urban areas, becoming the new binding element, the new relational matrix between what is urban and what is open space [4]. Understanding the meaning of this opportunity means giving way to new economies, new policies for social inclusion, and new projects for the landscape and for the rehabilitation of large settlement contexts. Significant are, in this sense, the French experiences, especially the "Ceinture Verte Régionale," which includes all of the municipalities in the Île-de-France region. The "Ceinture Verte," introduced at the same time as the "Plan Vert," part of the ***Schéma Directeur de la Région de l'Île-France*** in 1994, concerns an area heavily pressured by urban development, in which open spaces show increasing fragility because of the difficult transition between urban, peri-urban, and rural contexts. Peri-urban agricultural territories assume in this region the role of local development projects, whose management is entrusted to agro-urban programs aimed at establishing participatory processes between farmers and communities on issues related to the management, maintenance, and development of rural areas. These programs, along with specific policies for rural area development, have been generating new processes of economic, social, and environmental revitalization, that are increasing over time as different actions take consistency within individual local projects. It is a process that aims to restore the imbalances between the city and open spaces by enhancing the **multifunctionality** of agriculture into its material and symbolic components. This is a process capable of guaranteeing the development of this new rurality, the enhancement of local values and economies, and the construction of new forms of **contemporary landscape**.

European Community policies all turn to this direction, in all issues, starting from the "**Europe 2020 strategy**"[5], whose main aim is to stimulate sustainable, smart, and inclusive growth, in which local communities play a central role as actors and stakeholders[6]. The aims to be achieved and the risks to be avoided are now clear; planning is responsible for the implementation and for giving consistency to these innovative policies and strategies. The challenge we face is crucial, not only for environmental preservation, but mainly to ensure high levels of quality of life for the communities.

References

1. Acts of XII International Interdisciplinary Conference Volontà, libertà e necessità nella creazione del mosaico paesistico-culturale, Cividale del Friuli, 27/10/2007
2. Benedict M A, McMahon E T (2006) Green Infrastructure, Linking Landscapes and Communities. The Conservation Found. Island Press, Washington D.C.
3. Santolini R (2008) Paesaggio e sostenibilità: i servizi ecosistemici come nuova chiave di lettura della qualità del sistema d'area vasta. In: Riconquistare il Paesaggio, La Convenzione Europea del Paesaggio e la conservazione della biodiversità in Italia. Miur, WWF Italia, Rome
4. Donadieu P (2006) Campagne urbane. Una nuova proposta di paesaggio della città. Donzelli, Rome

[5] For further information, see: http://ec.europa.eu/europe2020/index_it.htm

[6] For further information, see: The Common Agricultural Policy after 2013 (http://ec. europa.eu/agriculture/cap-post-2013/communication/index_en.htm) and Technology & Innovation: Smart Cities and Communities (http://ec.europa.eu/energy/technology/initiatives/smart_ cities_en.htm)

8.4.6 Slow Paths and Urban Networks

Roberta Caprodossi and Paolo Santarelli

In the **reticular city** (or "territories network"), each pole can be defined as an intersection and exchange point within multiple flux networks joined by "infrastructure connection facets" that guarantee links for merchandise, people, information, and so on, in continual and reciprocal interchanges between the **local and global** dimensions [1].

Next to the settlement, productive and infrastructure networks, as well as digital and technological ones, environmental networks (including historical, cultural, and ecological connections) are imposed when characterizing and structuring the contemporary city, for reasons of environmental sustainability, landscape quality, the control of free land consumption, and to mitigate fragmentation.

Within this connective fabric, "**slow paths**" are found, a tight embedding of weak tracks that line the structure of the territory. These constitute a combination of short-lived boundaries of different types (**historical/cultural itineraries**, **pilgrimage routes**, forest/field paths, greenways, and so on) that cross heterogeneous geographical areas and relate to the specifics of local landscapes on different scales.

For at least a decade in many cities, considerable nonmotorized mobility projects have been initiated to increase the quality of life and the quality of the environment [2]. Today, these experiences are evolving toward planning solutions that are more integrated with the design of the urban space. The lengthy time needed provides an occasion to articulate projects for public spaces that are capable of introducing new natural aspects, socializing spaces, and connections between zones designated for different uses within the city [3].

The design of networks still responds to sectoral requests and has not formulated organic solutions for the territory. There is, therefore, strong selection between highly specialized zones, protected areas, and residual spaces, with all of their related environmental and social imbalances. "**Fast**" and "**slow**" are both essential conditions for the **quality of the urban space**, but functional proximity to different speeds is often translated into unsuitable, nonintegrated solutions. The complex system of relational networks—the "varied store of urban materials"—calls for a rethinking of the forms and uses of the city within a systemic framework to answer questions linked not only to climate change and diffuse accessibility, but also to those of habitability and inclusivity [4].

This represents fertile terrain for the transformation of the city—toward multifunctionality—to ensure the vitality of **multidimensional relationships** that will guarantee: the complexity of landscapes [5]; the development of technology on a human scale for new models of human/environment interactions; the realization of an urban interaction infrastructure through which citizens can participate directly to change services and the urban fabric; and the rethinking of open spaces, environmental, infrastructure, and settlement networks, as well as elements capable of contributing to reinforcing the identity of the contemporary city. In such a setting, slow paths can become the means for new scenarios to transform undefined spaces generated by the contemporary city, based on new design views of "networks and their interaction with the landscape" [6]. They cross spaces of different natures and scales: interstitial areas between built-up zones, residual spaces amid the networks, places in the local memory, borders of the still-farmed countryside; but also the edges of environmental corridors, waste or reuse areas, as well as new large technological parks. Their spatial–temporal dimension constitutes a connective armor, both physical and symbolic, that links cities and territories through spatial and anthropological relationships.

In the study of relationships between "naturality," "rurality," and "urbanity," characteristics peculiar to the reticular city, one can identify with the "*environmental question as a potent antidote to the fragmentation of the territory*" [7]. In addition to understanding the sense of the new multifunctional "**environmental infrastructure**", an understanding of the interactions between environmental networks and transformative dynamics of the city can open new doors to lend quality to the urban project.

The innovative conception of "slow paths" in a view of the reticular city can be a tool to deal with new spaces in the contemporary city, enrich them with a variety of uses and meanings, strengthen the link of belonging between inhabitants and places of everyday life, reduce degraded and abandoned areas, recover spaces that are usable on the human scale, and consider in an integrated way the complex of resources offered by the city.

References

1. Clementi A (2008) Corridoi, piattaforme, città senza fine. Nuovi scenari per la città medioadriatica. List-ActarD, Roma-Barcellona
2. Cities. Architecture and society, Architecture Biennale, Venice, 2006
3. Lotus International 139, Landscape infrastructures, Editoriale Lotus, Milan, 2009
4. Gasparrini C (2011) Una nuova città: diffusione e densificazione. Report to the Seminar "Ri-conoscrere e ri-progettare la città contemporanea". University of Camerino, Ascoli Piceno, 20 May 2011
5. Cassatella C (2011) Pianificare paesaggi multifunzionali: dalla scala vasta al progetto. Urbanistica 148, Inu Edizioni, Rome
6. Gasparrini C (2009) Nuovi racconti della città contemporanea. Urbanistica 140, Inu Edizioni, Rome
7. Gambino R (2011) I temi della transizione. Report to the Seminar Ri-conoscrere e ri-progettare la città contemporanea. University of Camerino, Ascoli Piceno, 20 May 2011

8.4.7 Relationships between Urban Green Spaces and Surrounding Nature Areas

Roberto Canullo, Giandiego Campetella, Stefano Chelli, and Marco Cervellini

The landscape collects both the presence of visible elements—natural and human—and invisible ones, such as ownership types, labor relations, and so on [1]; in other words, "landscape" means an area, as perceived by people, whose character is the result of the action and interaction of natural and/or human factors (European Landscape Convention, 2000).

The amount and availability of green space is strongly and positively associated with people's perceived general health than urbanity. Furthermore, this relationship occurs in different degrees [2]. However, urban and surrounding "**green areas**" differ in structure and function, providing different types of ecosystem services (and therefore, different levels of perceived quality of life).

The **Adriatic city** is characterized by a highly fragmented and heterogeneous landscape, where human infrastructures, urban green spaces, urban agriculture, and nature areas, penetrate each other generating a complex mosaic. This is particularly true for some of the most important coastal and **high-density urban cores** of the Marche Region (i.e., Ancona, Civitanova Marche, Pesaro, and San Benedetto del Tronto). This particular spatial pattern and its relationship with the perceived quality of life of the population who live and operate in this mosaic have not yet been studied.

In the future, and particularly for the Adriatic city, it seems fundamental to plan the urban and surrounding areas with a renewed methodology that considers the territory as a continuous dynamic system of ecological, human, and landscape components, in a more stable relationship. Coordinated and structured analytical approaches, like the one proposed here, can support decision-makers in informing spatial planning policies that maximize the beneficial effects of green spaces for the **quality of life** of urban residents.

8.4.7.1 Methodology

Considering the complex pattern of the landscape of the Adriatic city, our approach builds on a series of "green categories" [3] that are found to be associated with improvements in quality of life. This approach will be implemented through the integration of land use and land cover data (i.e., land cover classification and real vegetation maps) in a **geographic information system** database.

A spatial pattern analysis of the relationship between green patches and urban infrastructures will be performed using specialized **landscape analysis software** [4]. The outcome will help detect the organization of the mosaic and the relative scale at which variability and dependence of landscape units generate an optimum functional structure for the well-being of the residents [5].

The results of this first step are necessary to select particular case studies; hence, using a fine-scale approach, we should be able to: (1) quantify the surfaces and the distribution of each category (and subcategories); and (2) evaluate the spatial pattern structure and connections among categories through the use of dedicated landscape metrics [4].

According to previous spatial/ecological results, we will categorize the case studies into homogeneous groups, each one characterized by the same general spatial pattern of the categories previously mentioned. Using the nested classification, we aim to assess if there is a relationship between the types of spatial pattern and perceived quality of life. This will be achieved by integrating the geospatial characterization of the landscape mosaic with health data and structured and unstructured interviews with "residents" on the perceived quality of life [2, 3].

8.4.7.2 Results and Expected Impact on the Quality of Life

With this **multiscale study**, we expect to improve the knowledge of the Adriatic city, with specific regard to the spatial relationship among the most important green elements that determine the structure of urban and peri-urban landscapes. Furthermore, thanks to the integration of geospatial data on land cover and the assessment of the benefit of green space for quality of life through the residents' eyes, we expect to be able to integrate knowledge for policymakers that can have strong implications for **multiscale spatial planning**.

References

1. Ferrara G (1968) L'architettura del paesaggio italiano. Marsilio, Padua
2. Maas J, Verheij RA, Groenewegen PP, de Vries S, Spreeuwenberg P (2006) Green space, urbanity, and health: how strong is the relation? J Epidemiol Community Health 60: 587–592
3. De Vries S, Verheij RA, Groenewegen PP, Spreeuwenberg P (2003) Natural environments - healthy environments? An exploratory analysis of the relationship between green space and health. Environment and Planning 35, 10: 1717-1731
4. McGarigal K, Cushman SA, Ene E (2012) FRAGSTATS v4: Spatial Pattern Analysis Program for Categorical and Continuous Maps. Computer software program produced by the authors at the University of Massachusetts, Amherst.
5. Juhász-Nagy P, Podani J (1983) Information theory methods for the study of spatial processes and succession. Plant Ecology 51(3):129-140

8.5 Efficient and Nice

8.5.1 Settlement Dispersion and Urban Economies

Massimo Battaglia and Marco Frey

In economic terms, urban areas are characterized by the presence of externalities, as the result of the higher productivity that agents could achieve by being close to other producers or other market agents (so-called "agglomeration economies," in the form of ***urbanization economies***). In this sense, cities are the most economically efficient way for a spatial distribution of relationships among individuals. The basis for this higher efficiency is the existence of these "agglomeration economies," whose advantage has stimulated the phenomenon of urbanization in the last two decades of the 21st century. Agglomeration economies have been considered the key factor in offering higher incomes to the city households [1]. However, once a certain size is reached, ***agglomeration economies*** can degenerate into "agglomeration diseconomies," characterized by a progressive reduction of the advantages and positive effects of agglomeration [2]. In this perspective, in recent years, the attention has focused on urban sprawl and congestion phenomena, as manifestations of diseconomies at the urban level [3]. Capello and Camagni (2000) highlighted how literature identified the determinants of urban location rather than urban size or level of settlement dispersion, such as the type of economic function developed by the urban center, its spatial organization, and the efficiency of its internal structure. Moreover, recent studies have increasingly emphasized the role of other **qualitative determinants**, such as climatic variables, aesthetic elements, the presence of public goods and innovative services, effective local government policies, and social interactions [4–6].

This brief framework shows how, in addition to the **economic factors** that are important in explaining urban growth, a good quality of life (as a mix of good social relationships, environmental quality, attractive landscapes, excellence in government, innovative services, availability of public goods, and so on) represents a relevant factor for **sustainable urban growth**. These advantages constitute the "urban amenity," i.e., the desirable package of goods that can be demanded by the "consumers" at the urban level. According to the "urban planner" perspective, cities are the center of business, political life, commerce, and services: from this point of view, the analysis of the conditions, which contributes to the outline of urban amenity, is very important. Today this analysis can be developed into two different perspectives: (1) a perspective of applied research, based on specific urban contexts characterized by relevant growth; and (2) an operative–experimental perspective, oriented to design planning and local governance tools, whose aim is promoting the most effective and qualitative mix of "goods and services".

The first research perspective, in the different urban contexts, should favor the definition of the relative weights of the variables explaining quality of life with respect to the traditional determinants of urban growth, to measure the related contributions. This is needed to outline the potential connections existing between the specificities of the different urban contexts, their growth, and the values of the **quality of life indices**. The results would be classified into different groups using cluster analyses, giving the policymaker different (economic) situations among which to select the most effective strategies.

The second research perspective aims at providing specific operative tools of local governance and management to the public determinants to stimulate specific levels of development which can satisfy the **citizens' needs**, overcome the local conditions of potential disease, limit the pressures on the different environmental contexts, and favor processes of innovation and knowledge. This integrated approach needs continuous information detection systems at two different levels. First, it is essential to adopt and monitor an indicator system, which can involve the different dimensions of **local sustainable development** (in an economic, social, and environmental perspective). These **indicators** should be able to highlight the evolutions visible at the level of a certain urban area and to give policymakers suitable directions. This system should be accompanied by tools, which can collect directly the population's perception rather than the strategies adopted by policymakers. The detection tools can be several (permanent forums which involve selected stakeholders; the development of specialized focus groups among the citizens; surveys addressed to representative samples of the population via questionnaires). All these tools should provide the general vision that the population has of itself and the strategies that the public administration is realizing, showing the citizens' satisfaction because of the decisions taken.

The research perspectives mentioned previously can provide a valid tool for policymakers, who can use them to take decisions. On the one hand, according to the peculiarities of a specific context, it is possible to identify the dimensions of an appropriate development; while, on the other, through various territorial monitoring systems, it is possible to obtain indications revealing the effectiveness of the actions being undertaken.

References

1. Royuela V, Moreno R, Vaya E (2010) Influence of Quality of Life on Urban Growth: A case study of Barcelona, Spain. Regional Studies, 44, pp 551-567
2. Henderson J (1996) Ways to think about urban concentration: neoclassical urban systems vs. the new economic geography. International Regional Science Review, 19, pp 31-39
3. Camagni R, Gibelli MC, Rigamonti P (2002) Urban mobility and urban form: the social and environmental costs of different patterns of urban expansion. Ecological Economics, 40, pp 199–216

4. Glaeser E, Kolko J, Saiz A (2001) Consumer city. Journal of economic geography, 1, pp 27-50
5. Cheshire P, Magrini S (2006) Population growth in European city: weather matters – but only nationally. Regional Studies, 40, pp 23-37
6. Shapiro JM (2006) Smart cities: quality – of – life, productivity and the growth effects of human capital. Review of Economics and Statistics, 88, pp 324-335

8.5.2 Sustainable Mobility, Impacts, Quality of Life: an Economic Perspective

Luca Cetara

Economic reasoning can go far beyond its intuitive reach and help address the relationships between mobility and environmental policies, human development or quality of life.

A first group of economic readings of **sustainable mobility** are grounded on the concept of "costs," assuming that they can be discounted over time and substitution is possible between the costs borne in different periods. The possibility of covering the present and future external costs of mobility would assure sustainability [1]. Nevertheless, according to these definitions:

- The sustainability of mobility systems would depend on the estimation of future costs, which are by nature imperfect and not always possible.
- The perceived unsustainability of actual situations (which meet the requirements of the definitions) would be ignored.
- It would be difficult to suggest any active policy option, due to the focus on costs.

Thus, sustainability can hardly be identified looking only at costs. Other aspects have to be considered in a system, where sustainability results from a balance between layers. Analytically, this corresponds to defining a set of **indicators**, where all the layers of the system are represented [2].

A second group of economic readings highlights the economic relationship between mobility and regional economic growth: efficient mobility systems produce economic benefits that result in positive effects such as better access to markets and employment, increased productivity of businesses due to lower transport costs, and additional investment [3–6]; social benefits induced by transport planning policies (reduced commuting time lowering private defensive expenditures[7] and improving **quality of life** [2]); accessibility and affordability of services affecting **people's right to mobility**; availability of modal choice, affecting the quality of life of citi-

[7] Defensive expenditures are "*not directly sources of utility themselves but [...] regrettably necessary inputs to activities that may yield utility*" [8]. They "*are required to maintain consumption levels or the functioning of society*" [2]: e.g., expenditure on prisons and for commuting to work.

zens—notwithstanding some problems of measurability. Economic incentives in **mobility policies** can help increase efficiency by reducing unit costs (Euros per tonne mile or passenger trip) and favoring higher-value (emergency, freight, service, business trips, and high occupancy vehicles) over lower-value travel[8]. Efficient market principles (suitable consumer options, cost-based pricing, efficient prioritization, and neutral public policies, like congestion and pollution charges) can support economic development [6].

Nevertheless, external costs can overcome any benefits: inefficient transport systems lack in reliability or capacity and increase economic and social costs, with adverse effects on **regional welfare**. The environmental and social impacts of mobility systems are susceptible to economic valuation. They are critical in the presence of fragile ecosystems, vulnerable people, natural and cultural landscapes, for the link between regional traits, and opportunities for income generation (e.g., tourism [9]). **Social costs** include any impacts on health, employment, and income distribution [10], but also on land use, real estate values, and the incremental costs of induced trips [6]. Congestion and accidents also shape quality of life in terms of prospective capabilities, time consumption, and **health care**.

Decoupling the environmental impacts of transport from economic growth is a major challenge [11]. In wealthier countries, economic productivity tends to increase with reduced motor vehicle travel and higher fuel prices [6], when a "sustainability threshold" is broken. A balance has to be sought: mobility services have costs and benefits for the whole of society. Any exercise to address mobility and quality of life should include several cost categories and the influences that economics can help recognize and manage.

References

1. OECD (1996) Proceedings OECD International Conference Toward Sustainable Transportation. Hull
2. Stiglitz J E, Sen A, Fitoussi J P (2009) Report by the Commission on the Measurement of Economic Performance and Social Progress
3. World Bank (1996) Sustainable transport: priorities for policy reform. Washington, D C
4. World Bank (2008) Safe, clean, and affordable - transport for development. The World Bank Group's Transport Business Strategy 2008-2012, Washington, D C
5. Rodrigue J P, Comtois C, Slack B (2009) The geography of transport system. Routledge, New York
6. Litman (2010) Evaluating transportation economic development impacts. Transport Policy Institute, Victoria

[8] Some types of incentives can raise moral dilemmas, e.g., driving on the "express lanes" of US freeways by paying a fee [9].

7. Nordhaus W, Tobin J (1972) Is growth obsolete? Columbia University Press, New York
8. Sandel M (2012) What money can't buy. The moral limits of markets. Farrar Straus and Giroux, New York
9. Lanza A, Pigliaru F (1999) Why Are Tourism Countries Small and Fast-Growing? Working Paper CRENoS 1999/06, Cagliari
10. Greene D L, Jones D W, Delucchi M A (eds) (1997). The full costs and benefits of transportation: contributions to theory, method and management. Springer Verlag, New York
11. OECD (2006) Decoupling the environmental impacts of transport from economic growth. Paris

8.5.3 Innovative Transportation Systems and the Reduction of Traffic Congestion

Renato De Leone

In recent years, public transportation has faced two important challenges: to provide high-quality services that will satisfy the needs of end users and to maintain the same costs or even lower them. To deal with these two conflicting objectives, new **innovative transportation systems** have been devised and organized. By contrast, innovative transportation systems, such as **car sharing**, **carpooling,** and **dial-a-ride** (DAR) possess a high level of flexibility in responding to the requests of end users, and, at the same time, produce a significant cost reduction [1]. Moreover, the use of these systems also reduces **traffic congestion** and **environmental pollution**.

In a carpooling system [2], private cars are shared by a group of people who travel the same route. The benefits of carpooling are different depending on the various subjects considered. With regard to public administration, the main benefits are in a reduction of air pollution and traffic congestion. For companies or institutions promoting the system, the advantages are in offering a useful service to their employees, in a reduction of the parking areas, and in the consolidation of the corporate image. For the users, there is a clear monetary advantage in addition to special services that could be offered such as best parking lots, closer to their offices, or use of specific route lines.

In the car sharing system [3, 4], a car is used by several people. However, unlike carpooling, people use the car at different times and for different purposes. The car is owned by a public or private company. This type of service, while maintaining the advantages of a private car, allows a significant reduction of the fixed costs since the single member/user does not pay car property taxes or insurance, which are included in the car sharing service. Very often, the shared car is a small electric car, thus reducing environmental pollution. Moreover, users who car share may have additional benefits such as entering at any time in a controlled traffic zone or the use of dedicated parking lots.

In areas where transportation demand is weak, a possible solution to standard public transport services is represented by DAR services [5, 6]. This is a highly flexible service where the supply of transport is modeled according to actual transportation needs. In contrast to a standard bus service, where times and routes are completely fixed, in the DAR system, routes and times are recalculated on the basis of the users' requirements. Each request is received by a coordination/control center, which, based on the already collected and accepted demands and taking into account the availability of vehicles and drivers, determines whether the new request can be satisfied or not, and how to effect it.

It must be noted that all these innovative transportation systems require solving new challenging and interesting mathematical problems. Models for car sharing, carpooling and DAR systems have been proposed in the scientific literature and exact algorithms and efficient heuristic techniques have been studied.

All these innovative transportation systems require the presence of an efficient infomobility system that provides traffic and traveler information based on real-time data and geolocalization of the vehicles in the fleet. The expression **intelligent transport systems** refers to the set of equipment, methods, and procedures that allow, through data collection and acquisition, the processing of data and the dissemination of generated information to improve the transportation and mobility of people and goods. Vehicles equipped with specific terminals and **automatic vehicle location** or automated vehicle monitoring systems can provide up-to-date information on their position and speed as well information on traffic congestion, delays, and so on. This information, once properly processed, can be made available to public transportation managers and local administrators but also to end users.

Innovative transportation systems and infomobility are enabling technologies for **smart cities** [7]. The goal is to devise a decision support system that will integrate all the heterogeneous information sources concerning traffic and that will create and maintain a market for diversified mobility services, thus offering "smart services" and provide dynamic reallocation of mobility resources under special traffic conditions, thus offering "smart alternatives." This is a great opportunity but also a challenge for the next decade.

References

1. De Leone R, Di Girolamo N, Di Muro G, Marchitto E (2008) Pianificazione Programmazione e Gestione Innovativa per il Trasporto Pubblico locale. Aracne, Roma
2. Teal R F (1987) Carpooling: Who, how and why, Transportation Research Part A: General, Volume 21, Issue 3, pp 203-214

3. Benoît R, Dallaire Y (2007) Le DUO auto + bus :une alternative attrayante à la possession automobile, Routes & Transpor Vol.36, n 2, pp 34-36
4. Millard-Ball A, Murray G, Ter Schure J (2005) C. Fox. Car-Sharing: Where and How it Succeeds. Transit Cooperative research Program (TCRP) Report 108. Nelson\Nygaard Consulting Associates
5. Dial R B (1995) Autonomous Dial-A-Ride Transit Introductory Overview. Transportation Research C, n. 3, pp. 261-275
6. Colorni A, Righini G (2001) Modeling and optimizing dynamic Dial-A-Ride problems. International Transactions in Operational Research, 8:155–166
7. Di Muro G (2003) Analisi delle prestazioni di sistemi avanzati d'informazione all'utenza (ATIS). Tesi Sperimentale di Laurea in Ingegneria Gestionale, Università degli Studi di Roma Tor Vergata

8.5.4 Ecosystem Services and Urban Areas

Riccardo Santolini

In 2009, the percentage of the world population living in urban areas was over 50% and the United Nations [1] estimate that this quota will be over 70% by 2050. Furthermore, in industrial countries, demographic growth is not connected any longer to the availability of housing because this industry was assigned a driving role in economic growth regardless of the consumption of natural capital and of housing needs [2]. Indeed, different countries are experiencing different and complex phenomena of urbanization showing various degrees of degenerated peripheralization. In all of them, however, this evolution bears consequences for the **ecology** of (natural and nonnatural) ecosystems concerning both land use transformation and ecological functions loss, thus inducing **dystrophy** and increased **vulnerability**.

Irrespective of sunlight, urban areas are high energy-consuming entities with respect to **natural resources** (e.g., the use of agricultural products and water). All ecosystems are open systems powered by other ecosystems in a number of forms, energy, and information. This flow of energy characterizes the work of ecosystems, i.e., their capacity to provide goods and services (water and air quality, CO_2 absorption, soil protection, raw materials, recreational and cultural services, and so on) known as "ecosystem services" (ESS) [3].

Ecosystems, therefore, are mutually dependent, and urban areas are largely dependent on **natural capital**. The objective of bringing cities to less unbalanced and parasitic conditions can be a useful driver for the adoption of best practices and interesting actions. Nowadays, innovation involves understanding the need for the adoption of an ample and systemic perspective when considering a landscape and its flows of energy and materials (i.e., food, water, energy), which feed a urban ecosystem. A **systemic perspective** needs to include the ecosystem that receives the waste produced by the urban one (i.e., the areas where solid, liquid, and gaseous waste is conveyed).

Thus, the goal is (at the very least) to reduce parasitism by showing how cities can reduce their demand of materials and energy by not only reducing consumption and avoiding wastage, but also by leveraging opportunities for the cities to become themselves "producers" of materials and energy. One key factor is the recovery of sunlight energy systems within cities and in nearby rural areas. Such systems must be a subsidiary yet integral part of an urban system; they have to be planned and designed specifically because they are part of the natural capital that cannot be substituted by other kinds of capital, nor can they be compensated for. In fact, the natural urban ecosystems which produce ESS contribute to public health and increase the quality of life of citizens since they improve their living conditions and increase the economic value of the relevant urban area.

The value of the functions of these ecosystems is considerable: Elmqvist and coworkers [4], drawing on a number of works (ten studies concerning nine cities in China and one in the USA), showed the monetary value of seven ESS provided by green urban areas. On average, recreation, amenity, and health effects are valuated at USD$ 5,882 and USD$ 17,548 ha/year.

City governments are increasingly responsible for providing services to the resident citizens. Since they are working at the grassroots level, they are the stronger supporters of change. Awareness-building concerning the dependency of urban areas on healthy environments that produce a stable supply of ESS and offer opportunities for employment (ESS need maintenance), is the road to real sustainability. When discussing possible ways to transform local areas, effective communication can be built by prompting discussions on natural capital and **ESS** so that the value of local tangible and intangible resources and the meaning of environmental impact can be more clearly conveyed. This way, the benefits deriving from a resilient environment and the ESS appear more evident to both citizens and decision-makers. Any threat to the ecosystem will then have to be considered according to the **economic environmental balance**, in terms of both the benefits that would be lost and the social and economic damages caused. The study of large urban areas, such as ecological systems, tends to recover the equilibrium between the city and its external environment in the exchange of **energy** and information. This process is long and complex (and perhaps unrealistic). Rather, it is important to design (and then manage) the buildings, the urban topography, the green space systems, and so on, as appropriate components of an integrated planning that takes adequately into account the ecosystem and its cycles (water, energy, waste). This approach, in which they are regarded ecosystem functions and the economic valuation of ESS as structural elements of planning, is really a strategic objective to be aimed for as they promote the durable development of cities and the well-being of their citizens.

References

1. UNHABITAT (2008) State of the World's Cities, 2010/2012 – brinding the urban divide. Earthscan, London, pp 222
2. Liu J, Daily GC, Ehrlich PR, Luck GW (2003) Effects of household dynamics on resource consumption and biodiversity. Nature 421: 530-533
3. De Groot RS, Wilson MA, Boumans RMJ (2002) A typology for the classification, description and valuation of ecosystem functions, goods and services. Ecological Economics 41:393–408
4. Elmqvist T, Setälä H, Handel S, van der Ploeg S and de Groot R (2011) Benefits of ecosystem services in cities. Article to be submitted formally. In: TEEB, The Economics of Ecosystems and Biodiversity (2011). TEEB Manual for Cities: Ecosystem Services in Urban Management. www.teebweb.org

8.5.5 Form of the City, Urban Sustainability, and Landscape Quality

Rosalba D'Onofrio

The settlement model known as "**urban sprawl**", which has been proposed in recent decades, in addition to posing questions about the overall design of the city, the definition of its limits, and the recognizability of the **landscape**, is calling the attention of experts, scholars, politicians, and private citizens to its effects on the urban ecosystem. In particular, questions are being raised about the consequences of urban sprawl on the splintering of natural habitats; the compromise, simplification, and homogenization of species and the landscape; the alteration of **hydrogeological systems**; and the modification **of energy fluxes**. In addition, the questions ask how these changes influence the capacity of ecosystems to sustain human functions, the well-being of the city's inhabitants, and their **quality of life**.

In this new perspective, the objectives of sustainable development and the livability of the urban environment reinforce each other to the point of hypothesizing that, to reach the general objective of a livable future in urban areas, the design of the city itself must change. To date, the effects of different **shapes of the city**, different densities, and different levels of urban connectivity on the functioning of the ecosystems are not known; nor is it known the way in which interactions between these different components vary from place to place. For about 15 years, groups of international researchers have been investigating this theme on different spatial scales. On the **regional and metropolitan scale**, we bring to attention, for example, the studies by Forman [1]; on the urban scale, those by Girardet, Nijkamp and Perrels [2], and others; and on the neighborhood scale, those by Corbett and Corbett [3], Rudin, and Falk. Despite some attempts [4], there is no common conceptual framework today that allows one to compare these different approaches and consequently to move from them to the identification of a desirable urban form and the consequent urban politics. In short, one feels that a theory is lacking to help evaluate if one specific urban form can contribute more than the others to making the contemporary city more sustainable.

8.5.5.1 Progress of the Disciplinary Debate

The interdependence between anthropic and natural processes poses unprecedented challenges in managing the growth and requalification of cities, and requires a new frame of reference to guide the evolution of an urban shape that is capable of better responding to the complex interactions between human processes, environmental and **ecological resilience**, and the landscape. However, this does not mean identifying an "ideal urban form" that is completely adapted to different ecological processes or that totally ensures the conservation of biodiversity and the quality of life. It is important to note that the urban environment is much more variable and complex than has been conceptualized already; in real systems there are multiple equilibriums and the best response can change under different conditions, as well as evolving in time. However, since different urban models at different scales deal with different grades of compactness, density, connectivity, and heterogeneity of the urban fabric, it is worth asking what the best combination/interaction may be to guarantee a reasonable equilibrium. In particular, it should be understood how:

- Models of land use, constructed environment, and open spaces interact with the energy system of the city and how the structure and spatial form influence the **energy needs** of a series of activities, especially transportation and the heating of manufactured buildings.
- Different settlement shapes have different effects on biophysical processes, the ecological conditions of the area, environmental comfort itself, and the conservation and improvement of the landscape.

To do this with the aim of testing a new methodology to evaluate the urban form that is aimed at sustainability, the following are necessary: a comprehensive **dynamic analysis** of urban systems with the support of multiple integrated disciplinary skills and recourse to a parameterization model that is capable of identifying an equilibrium point among possible action scenarios for the development of these territories, the requalification of the city, and the sustainability objectives identified.

8.5.5.2 Expected Results

A methodology designed in such a way should be able to supply public administrations, designers, and planners with a tool to evaluate the implications of alternative urban development models within the goal of sustainability. In addition, it should provide valid support for decisions when defining alternative development scenarios, identifying urban politics, planning choices, and integrated and shared designs, and sustainably developing the landscape.

References

1. Forman R (1997) Land mosaics: The ecology of landscapes and regions. Cambridge University Press, Cambridge
2. Nijkamp P, Perrels A (1994) Sustainable cities in Europe. Earthscan, London
3. Corbett J, Corbett. M (2000) Designing sustainable communities: Learning from village homes. Island Press, Washington, DC
4. Williams K, Burton E, and Jenks M (2000) Achieving sustainable urban form. Routledge, London

8.5.6 Architecture, Technologies, and Urban Quality

Federica Ottone

The technological dimension of architecture has to do primarily with the capacity to express innovation through the forms of language typical to architecture: thus, **architectural innovation** that is expressed not through linguistic syntax, but rather through forms, elements, and devices dictated by the necessity to track the evolution of external conditions, of necessities, of instances deriving from changes to social, economic, and productive conditions.

Within this framework, for some time now, the themes of **renewable energy** and energy savings have come to occupy a relevant position in both research and the governance of processes transforming the city and the landscape. However, today, the greatest urgency is that of intervening in the existing city, seeking to remedy the shortcomings in the overall quality of the **built environment**, beginning with the quality of buildings realized between the 1950s and 1980s[9]. The urgent need to intervene is, however, discouraged by the incapacity to begin a reflection on the economic implications of an eventual technological and **architectural reconfiguration** of public and private building stock.

This acceptance of responsibility implies the identification of concrete examples of experimentation to effectively establish to what degree the advantages of an action of energy refurbishment may be supported by soci-

[9] Italian Law n. 10/91: Regulations for the Implementation of the National Energy Plan for the Rational Use of Energy, Energy Savings and the Development of Renewable Energy Sources, disciplined by implementation regulations only in 1993 and 1999. These years were marked by the development of a greater awareness of issues related to energy savings.

ety and public administrations, and eventually what interventions can be considered most sustainable, after placing them in relation to other hypotheses of an entirely different nature.

Are we sure, for example, that the costs of an energy retrofit to an existing building, which in Italy are in the range of one-third of the cost of new construction[10], are effectively sustainable if we add the costs of retrofitting the same buildings to resist seismic action?

In many cases it is necessary to establish points of equilibrium between disciplines within the same macrodisciplinary field. However, often, as this volume attempts to demonstrate, there is a tendency in particular choices to omit a number of determinant implications deriving from priorities dictated by diverse scientific fields, to establish the convenience or inconvenience of an action of **building regeneration**.

When we pass from the scale of the building to that of the city, each single intervention must be able to condition and offer benefits to society in general.

We must identify forms of sustainable intervention, for which there exists a tool suitable to evaluate their effectiveness in relation to the diverse conditions present in a given territory.

An interesting pilot project has been realized in the Portuguese city of Porto. The **Arrebita! project** proposes the activation of a process for the restoration of an abandoned part of the city through forms of participative action, bringing together various stakeholders and working with economic sustainability ("*a collaborative system that allows for the refurbishment of derelict buildings free of cost*")[11].

Public administrations consider the need to refurbish abandoned parts of the city a primary objective; this is accompanied by aspects of energy savings, now both urgent and unavoidable that, coupled with the current economic crisis, render any form of project financing difficult to believe in [1].

In Europe, experiences of **energy retrofitting** involving public buildings, almost always the object of European financing, are currently being evaluated to verify the relationships between costs and benefits in terms of **energy savings**. One example involves a large district in East Berlin, entirely refurbished and the object of significant financial investments[12]. In both cases, even though very different, we are witness to a prevalence of the

[10] One example in support of these data is the energy retrofitting of the Mathematics building of the School of Science and Technology at the University of Camerino, designed by the spin-off PEnSY, at a cost of approximately 350 Euros/m^2.

[11] Arrebita! Porto is the social entrepreneurship project winner of the competition "FAZ— Ideias de Origem Portuguesa" organized by two major philanthropic foundations in Portugal, the Fundação Calouste Gulbenkian and the Fundação Talento. The aim of the competition was to summon the often neglected expatriate community and challenge it to come up with ideas for the development of their homeland (http://www.arrebita.org/files/Arrebita_info_EN.pdf).

functional and technological program and decision-making process, relegating the formal architectural results of each single intervention to a minor role.

It is necessary to understand whether an architectural project, which may be considered "weak" in offering services under a wide variety of conditions, is instead capable of constructing a diverse and new range of more useful and lasting values, as a function of the improvement of the quality of life and the built environment.

References

1. Harvey LDD (2009) Reducing energy use in the buildings sector: measures, costs, and examples, Springer Science + Business Media BV. DOI 10.1007/s12053-009-9041-2

8.5.7 Renewable Energy, Thermal Comfort of Open Spaces, and Environmental Quality: Management of a Dynamic Equilibrium

Roberta Cocci Grifoni

Research on **environmental comfort**, in particular as linked to open spaces and the design of public places in the city, is a theme that is still not dealt with much in the literature. With increased environmental sensitivity, an interest in improving the comfort of open spaces has recently been witnessed; it is a theme closely connected to the politics of environmental sustainability, the reduction of energy consumption, and greenhouse gases, as evidenced by the growing attention to **climate change** in recent years.

Such research has always been substantially based on the consolidated studies conducted in the indoor environment [1]; however, it has now been demonstrated how this research cannot be applied to open spaces, since indoor discomfort is much more controlled and resolved through **engineering/installations**. In the outdoor case, however, the correct design of urban public spaces is necessary.

Environmental comfort can be developed starting from two different but coplanar points of view, i.e., the well-being of people, linked to **quality of life**, and the containment of energetic resources, linked to the sustainabili-

[12] One of the most important experiences of energy retrofitting in Europe is in the Märkisches Viertel, Reinickendorf, Berlin. See Berliner KlimaSchutzPartner, EUMB Pöschk/VME – Verlag und Medienservice Energie (2010) Climate protection - Good projects from Berlin - Senatsverwaltung für Gesundheit, Umwelt und Verbraucherschutz – Öffentlichkeitsarbeit.

ty and use of renewable resources. Some studies, arising from apparently distant disciplines, such as sociology [2], have demonstrated the link between quality of life and well-being, studying citizens' way of life in the principal European capitals in relation to various parameters, among them the use of urban open spaces and the environmental qualities present. Also interesting is the theme of sustainability: the gradual substitution of nonrenewable resources with renewable ones takes on essential value for the protection of our living environment and for air quality.

In the contemporary city, however, open space is very often represented by a polluted urban "void," similar to a "non-place," according to the definition by Mark Augè [3], which defines the connective fabric of the city and the cohesive element of the urban landscape.

8.5.7.1 Advances in the Disciplinary Debate

A characteristic that allows an open space to reach the definition of "place" is the quality of the space. The space becomes a place in virtue of perceived well-being, in virtue of appropriately researched and designed environmental comfort. The way in which we recognize and experience the space ensures that we attribute to it a particular significance that transforms it into a place with **air quality**, where atmospheric pollution has been reduced to a minimum.

The level of quality of life is orienting itself toward ever higher standards, and it thus requires the extremely careful and precise definition of microclimate and air quality requirements capable of guaranteeing **thermal/moisture comfort** (globally and locally) in the spaces where people live and work, and which are necessarily related to energy savings and environmental sustainability.

Since the place is designed around and characterized by values of temperature, pressure, and relative humidity, it is necessary to consider it as a thermodynamic system to design it correctly.

Starting from these definitions, the design of a place can be considered as the "thermodynamic mediation" between the body and the space (environment), between movement and quiet, between meteorology and physiology [4]. This oscillation between indefinite space and the concrete place of habits is the stage on which an architectural and urban planning project that wants to reach a quality-of-life objective should move.

The human organism, just like the urban organism, through human and urban metabolism phenomena respectively, produces thermal energy that disperses in the environment through different heat transfer mechanisms. In resting conditions, human or environmental temperature can vary by a few degrees and to maintain such a state, it is therefore necessary that the energy produced through energetic/metabolic processes equals the energy dispersed.

Recognizing the importance of such analysis, it is necessary to begin working on thermal balances, the analysis of fluxes in the environmental context, meteorology, **polluting emissions**, the **meteo-climatic characteristics of a place**, and the relationship between buildings and open space using a systematic approach. It is necessary to recognize that the space to be designed includes a complex combination of factors that may be energetic, environmental, landscape, economic, social, and so on. New forms of the project should grow from attempts to support the energetic vocation of a territory, to favor climatic forcing, to reduce climate-altering emissions, and to respect the landscape with the goal of achieving quality of life within continually changing external environments.

8.5.7.2 Expected Results

The goal of this approach is extended to the planning of quality urban spaces linked to the awareness that improving the quality of the landscape and therefore its environmental conditions will allow for the creation of comfortable open spaces characterized by a better quality of life, in addition to the encouragement of social relationships between people. Therefore, through an **"integrated" plan**, a design characterized by a complex of intersectoral actions that are coherent and closely interconnected, one can converge on the common goal of environmental **sustainability**, thus one of the main objectives in environmental design in the urban context.

References

1. Fanger P O (1970) Thermal comfort. Analysis and applications in environmental engineering. McGrawHill, New York
2. Nuvolati G (2007) Commuting and Quality of Life: The Italian Case. In: Estes R (ed) Advancing Quality of Life in a Turbulent World. Springer, New York
3. Augé M (2009) Nonluoghi. Introduzione a una antropologia della surmodernità. Eleuthera, Milan
4. Rahm P (2009) Architecture météorologique. Archibooks, Paris

8.6 Clean and Healthy

8.6.1 Waste Prevention and Improvement of Citizens' Quality of Life in Terms of Socio-Economic and Environmental Health

Barbara Fenni and Sara Spuntarelli

The "**waste** problem" is linked to the Industrial Revolution, the change from an agricultural socio-economic system to a highly industrialized one with a strong environmental impact. The situation became even more complex after the Second World War, when low-cost, poor quality goods were produced, sold, and disposed of after a short time. Waste production was justified as inevitable with the increase of gross domestic product (GDP) and its management was limited to dumping and incineration. Italy's main problem was a lack of integration between territorial and refuse management planning. Emergency situations in our country placed the landscape into a secondary place in the selection of disposal sites. The choices affected valuable open spaces and regions close to inhabited centers, creating many concerns for the populations involved. The scientific community confirmed the need to acquire more data, noticing a link between the increase of some types of **cancer** and malformations and the closeness of disposal sites, both for incineration [1] and dumping [2]. Additional problems included the constant flow of motor vehicles carrying waste, the region's decrease in value, the negative impact on farming and tourism, and the elevated costs sustained by citizens. At the beginning of the 1970s, the international community became more sensitive toward the environment, highlighting the need to curb the population growth rate, industry, pollution, and the exploitation of resources to avoid exceeding the **planet's "load capacity"** [3].

In the "**Sixth Community Environment Action Programme**" of the European Community (EU), 2002–2012, the EU stressed the need to develop specific measures of prevention and recycling as a priority to stop the exponential growth of waste[13]. The term **prevention** refers to actions adopted before a substance becomes waste and determines quantity reduction[14], danger, and impact on the environment and **human health** [4]. Also important is separate collection and the successive recycling of the subdivided materials allowing the reintroduction of waste into the economic cycle in the form of quality products to avoid the use of other resources.

[13] Decision no. 1600/2002/EC of the European Parliament and Council which instituted the Sixth Community Environment Action Programme, http://eurlex.europa.eu/LexUriServ/LexUriServ.do?uri=CELEX:32002D1600:en:NOT.

[14] Art. 3, point 12, letters a), b), and c) of the European Directive 2008/98/EC, http://eurlex.europa.eu/LexUriServ/LexUriServ.do?uri=CELEX:32008L0098:EN:NOT.

The importance of these actions for waste management has been received in different ways within the EU: among the most virtuous are Germany, where 70% of waste is **recycled**, Sweden, the European leader in the Waste Electrical and Electronic Equipment directive, and Norway where rubbish acquires real value, as many supermarkets have automatic waste collectors for tins and plastic bottles. In exchange, consumers receive shopping vouchers. Instead, in the last EU report dated 2 July 2012 on correct waste management in the 27 member countries, Italy was only in 20th place, as prevention is not a national concept but limited to single initiatives [5]. **Separate collection** also moves at two speeds: in the north, it is around 60%, but in the south, it is still under 30%. Of the 95% of waste dealt with, more than half continues to be dumped, 15% is incinerated, and only 34% is recycled or made into compost.

This data highlights the need to put waste prevention into practice to defend and guarantee the right of **environmental health**, a collective social right considering the relatively recent destructive process of the planet. The fundamental rights are historically and culturally determined values and the vital needs of each person. In this sense, the right to environmental health and quality of life of citizens are fundamental rights and must be guaranteed by the State [6]. In the immediate future, Italy must adopt ways of discouraging dumping and incineration, which is hazardous to health and the environment, and support prevention and separate collection. Proposals include introducing taxes and bans on traditional dumps and incinerators and campaigns to awaken concern in schools and workplaces. Producers should be more responsible and rethink their **production strategies,** while consumers must change their own **waste habits**. These are necessary strategies to restore value to the countryside for citizens to enjoy and live healthily.

References

1. Cordier S (2010) Maternal residence near municipal waste incinerators and the risk of urinary tract birth defect. Occup. Environ med. 2010, 67 493-9, http://oem.bmj.com/content/67/7/493.abstract
2. World Health Organization (2007) Population health and waste management: scientific data and policy options. Report of a WHO workshop, Rome, Italy, 29-30 March 2007, http://www.euro.who.int/document/E91021.pdf
3. Meadows D H, Meadows D L, Randers J, Behrens III W W (1972) Rapporto sui limiti allo sviluppo. Commissioned by the Club of Rome al MIT, Mondadori
4. Martino R (2010) Compendio normativo per la gestione dei rifiuti e la bonifica dei siti inquinati. Maggioli, Napoli
5. European Commision (2012) Screening of waste management performance of EU Member States. http://ec.europa.eu/environment/waste/studies/pdf/Screening_report.pdf
6. Ferrajoli L (1989) Diritto e Ragione. Laterza, Roma-Bari

8.6.2 Detailed Climatic Analyses for Optimal Territorial Governance and Sustainable Development

Carlo Bisci

Environmental parameters and quality of life are strictly interrelated and both strongly depend, directly and indirectly, on local meteorological and climatic features. The latter, in turn, have been significantly altered by recent complex **global climatic changes**, making it even more difficult to create a detailed model of their spatial and seasonal variation, and therefore to properly delineate a local development plan and to follow best practice for optimal territorial governance.

Therefore, the primary aim of research in applied climatology generally is to identify a circumstantial pattern of spatial and seasonal distribution of the main **climatic parameters** and **bioclimatic indices** in the study area, while also evaluating from a probabilistic point of view their evolutionary trends in the short and medium term (i.e., for the next few decades). Only by working from these data, in fact, it is possible to give sound information regarding the potential impacts (positive and negative) of the local micro- and topoclimates (and of their variability and variation) on land use suitability and on natural hazard and risk, which are instrumental to adequately interpret natural environments as well as social and economic problems for proper land planning and territorial management.

Present knowledge of global climatic features is good, particularly in European countries where a fairly dense network of recording stations exists, such as in Italy (where, except for some mountain ranges, their distribution is generally good for precipitation data and adequate for thermometric ones). On the other hand, **detailed analyses of topo- and microclimates** that are instrumental for land use planning and territorial governance are sporadic and extremely inhomogeneous.

It is therefore fundamental to carry out research to define models to quantitatively describe the spatial and seasonal distribution of the various parameters. These studies should be based on the integration of historical climatic records with environmental data, such as morphometry, elevation, slope angle and aspect, albedo, land use, vegetation, valley and ridge trends, and so on. In fact, only by analyzing synergistically all the above information, it is possible to produce reliable and detailed maps of the climatic features and indices that are needed for optimal territorial planning and governance.

For present climatic change, the reports and technical papers of the **Intergovernmental Panel on Climatic Change** are generally considered as the most reliable reference on a global scale. Starting from them, it is possible to downscale the results to local situations, analyzing nonperiodic time variations of climatic records to identify variations and to give quantitative evaluation of their levels; in this way, it is possible to extrapolate the evolutionary trends of the phenomena and, most of all, to provide scenarios of the environmental variations that are likely to happen in a few decades.

To obtain such results, first of all it is necessary to collect the climatic and environmental data available and to submit them systematically to critical analysis, homogenization, and georeferentiation before using them to feed a properly designed, multidisciplinary geodatabase. The use of advanced **geographic information systems** is instrumental not only to manage the vast amount of data but also to perform simulations, to calculate ancillary and derivate layers of information, to carry out statistical and geostatistical analyses, and to produce thematic maps. All the climatic records collected at every station should be thoroughly analyzed to obtain the statistical parameters that are needed to describe the **local climate**, as well as to identify both cyclic and nonperiodic variations. The latter constitute the basis for the elaboration of a mathematical model describing future local climatic trends. In parallel, environmental data (physical and biological) should be used to create, test, and calibrate a complex algorithmic model of the spatial distribution and variability of the main features influencing micro- and topoclimates. This model, in turn, is applied to climatic data to regionalize them, thus obtaining detailed maps that quantitatively depict their spatial variation. The same algorithms should finally be applied to the forecast values of climatic parameters and **bioclimatic indices** to produce maps of the probable scenarios, and to describe the spatial distribution of the probability of any given climatic variation.

Detailed information on the spatial and time distribution and variability of climatic parameters and bioclimatic indices, when properly cross-refenced with the other relevant information, make it possible to adopt optimal strategies for territorial governance as well as to draft the best possible plans for sustainable development and the exploitation of environment and territory. In particular, it makes it easier and more effectual to delimit and characterize areas prone to natural hazard (also quantifying the related risk, thus reducing the overall average cost paid for safety and remediation), to choose adequate agricultural practices as a function of local land use suitability features (making it easier to maximize both land profitability and sustainability), and to plan and realize artifacts and interventions that are more adequate to present and future needs and situations, to achieve the maximum result with the minimum possible effort and impact, and to obtain an actual improvement in quality of life and landscape.

8.6.3 Geomorphology of Places and Hydrogeological Risk

Bernardino Gentili

The presence, quality, and quantity of numerous natural resources are closely linked to the geological characteristics of the territory, but so too are different conditions of geoenvironmental hazard, which are connected to high-level risks. These depend: (1) directly on the geodynamic evolution of the Earth's crust, as in the case of seismic and volcanic risk ("geological risk"); and (2) on the interaction between geological and climatic conditions, as in the case of "**hydrogeological risk**."

Geological studies have always given precedence to analyses concerning natural environments and, as a rule, have more or less avoided intensely humanized ones, particularly urban areas, because man has changed the original structure, at least as regards the most superficial levels of the Earth's crust.

However, given that there is no work or human activity that does not have relationships (more or less close) with rocks and earth, in recent times (particularly in the last three decades), the need to extend geological studies even to urban areas, with emphasis on aspects related to application, has been felt with gradually increasing intensity. Specific research programs aimed at adjusting operational methods have been developed to apply to: (1) the definition of the natural geological model of the urban substrate; and (2) the recognition of the changes introduced by humans in the **geological structure** and the correct definition of their impact on the physical environment [1, 2].

8.6.3.1 Hydrogeological Risk

In this area, the research on the mass movements that in "young" countries, geologically speaking (e.g., Italy), are responsible for the largest number of casualties and environmental damage after earthquakes, is of primary importance.

Historical urban centers located on the top of mountains appear to be particularly vulnerable to mass movements; this situation is far more common in territories of older urbanization, because they are "attacked" by the upstream regression of the phenomena which are generally activated at lower altitudes; the instabilities are generally slow or very slow-moving. This is not the situation of towns located at the foot of the slopes or in an intermediate position: they are generally subject to the mainly disastrous impact of fast-moving masses.

Other hazards of noteworthy importance are linked to the action of surface water and groundwater. The first is often associated with fluvio-torrential processes, which in urban areas located at the foot of a slope or in a plain give rise, respectively, to prevalent phenomena of **debris flow** or **flood**. These effects are associated with the following: large reductions in density and consequently of the draining sections of the natural **river hydrographic network**; the incorrect sizing of the openings of the bridges and drainage tunnels; excessive and/or "casual" use of subways that are almost devoid of adequate drainage and underground works designed to protect them from the surface water affecting them; the creation of reliefs that slow down (or prevent) natural surface drainage of water and which, furthermore, promote the deposition of sandy-loamy materials transported by them, thus waterproofing soils, with the consequent elimination of underground water drainage; and the almost systematic lack of maintenance of drainage systems.

To the underground cavities, either natural or of anthropogenic origin, are often associated phenomena of failure or collapse of artifacts. The presence of natural underground cavities, typical of rocks sensitive to karst processes, may be assumed from the careful analysis of a detailed geological map, while the presence of anthropogenic underground cavities, achievable on different lithotypes, can be known only through oral or historical testimony.

Significant progress has been made on the knowledge of the activation and evolutionary control factors of the different types of hazard mentioned previously, allowing the zoning of geoenvironmental hazards to be carried out with considerable confidence, but there are still unresolved problems related to the temporal prediction of events.

The technique is different depending on the type of **risk**, and always provides a deeply detailed geological-technical study of the territory, supported by historical, geophysical, and geomechanical investigations and by possible installation of monitoring systems. The purpose is to check whether the anthropogenic (or other) "structure" falls within the "firing range" of the process under consideration or in areas where it can significantly feel its effects.

To do this, we must have a thorough knowledge of the lithotechnical characteristics of the land forming the foundation of the structure and the **hydrogeomorphological characteristics** of the site and of an adequate surrounding area. With a good approximation of the magnitude of the phenomenon, it is possible to move on to engineering analyses, designed to assess the likelihood that the anthropic structure fulfills the necessary requirements to withstand the stress caused by the process.

The analysis of geoenvironmental risks is both essential and a priority for any action of spatial planning aimed at improving the socio-economic conditions of citizens, for whom **geoenvironmental security** is of "vital" importance.

References

1. Aringoli D, Gentili B, Materazzi M, Pambianchi G (2010) Mass Movements in Adriatic Central Italy: Activation and Evolutive Control Factors. In: Werner ED, et al (eds) Landslides: Causes, Types and Effects. Nova Science Publishers, New York, pp 1-71
2. Huggenberger P, Epting J (eds) (2011) Urban Geology. Process-Oriented Concepts for Adaptive and Integrated Resource Management. Springer, Basel N° XVI

8.6.4 Coastal Dynamics and the Naturality of Littorals

Alessio Acciarri, Carlo Bisci, Gino Cantalamessa and Giorgio Di Pancrazio

Coastal dynamics issues would not have had particular relevance a few decades ago, when attention was focused on the advance of beaches and the silting up of harbors more than on **littoral erosion**, and when it was a common belief that human actions could not effectively counteract natural morphoevolutionary trends, such as the reduction of relief and shoreline progradation.

After the Second World War, the situation began to change and the shoreline trend inverted in the 1970s, when strong coastal erosion started to threaten many facilities, such as roads and railways, and seaside resorts were heavily damaged, dramatically compromising touristic and seashore activities.

According to most scientists, the decrease in solid load brought to the sea by rivers is indeed the primary cause of such erosion, at least in Italy. This sharp and dramatic reduction of sediment supply to the beaches has been caused by extraction of material from riverbeds, obstruction of rivers by artifacts, such as dams and river cross bars, the abandonment of fields, and the reforestation practices of mountain slopes, all connected with industrialization and strong urbanization of plains and coastal areas. Moreover, **climate change** has been causing a continuous rise in sea levels and more frequent storms with higher risk of seawater ingression in the coastal plains and the destruction of characteristic dune environments (e.g., mouth of the Po River, the Sentina Regional Natural Reserve, and so on).

This situation fully reflects what is now happening along the coast of the **Marche Region** (and along almost every portion of the Italian shoreline). Recently, local and national newspapers have frequently reported of extensive storm damage to beaches and facilities. As a consequence of this regressional trend, many expensive counteractions, mostly funded by local governments, were adopted. Very often, such **coastal defense works** (e.g., breakwaters, groins, and so on), lacking adequate planning to prevent negative effects on the coastal dynamics of the site and of neighboring areas, have been the cause of further **erosion processes** in the surrounding, unprotected areas. As a consequence, about two-thirds (116 of 172 km) of the Marche regional coast is characterized by the presence of hard coastal defense works that are often causing **environmental degradation**.

Therefore, while also taking into account the importance of beaches in the economy and quality of life of coastal areas, it is vital to study their evolution and identify optimal techniques to protect them. In fact, only working from detailed multidisciplinary research—including the seasonal monitoring of mineralogical, morphological, and sedimentological features, as well as studies of meteorological and marine features, and coastal hydraulics—carried out both on the submerged and emerged beach, will it be possible to find fully adequate countermeasures to the problem. Small differences in wave climate and/or in the features of a remedial work may induce a break in shore equilibrium leading to completely new morphologies, thus heavily influencing the anthropic activities located in the vicinity.

Comparing the results of continuous monitoring of littoral evolution (taking into account the morphological, bathymetric, sedimentological, and mineralogical properties of backshores and shorefaces) with local wave climate and meteorological parameters makes it possible to forecast future scenarios and trends. Such studies are instrumental first of all to understand in detail the duration, intensity, and typology of the sequence of erosional and depositional cycles determining the local littoral evolution. Working on the basis of these studies, it will be possible to plan and carry out adequate countermeasures to solve the problems without inducing potentially dangerous side effects, and to formulate optimal local and regional interventions, setting up integrated and sustainable governance plans. This point of view is supported by the European Union, which in Recommendation 2002/413/EC invites member States to adopt a strategic approach aimed at environmental, social, and economic sustainability for the management of coastal areas.

The proposed integrated and multidisciplinary research on basic knowledge about the shores of the "**Adriatic Region**" may be instrumental in providing stakeholders (local governments, "Marche Region Agency for Environmental Protection," "Italian Agency for Environmental Protection," touristic operators, and so on) with the appropriate framework for optimal governance of the littoral environment and territory. As a consequence, scientific work positively impacts the economic development, unemployment, and quality of life of **coastal towns** in:

- Tourism, leisure, and landscape: improvement of the touristic use of beaches and sea through conservation and reclamation.
- Fishing: diversification of professional and sport fishing activity based on the recognition of the spatial and temporal distribution of different marine environments.

- Agriculture and industry: the protection of littoral areas from erosion will allow better planning for the distribution of productive activities as well as obviate problems with saltwater intrusion into water wells.
- Transportation and trade: blocking the sea from advancing will prevent instability risks for railways and roads, as well as reduce silting problems in harbors and marinas, thus favoring touristic and commercial activities.

8.6.5 Indicators for Air and Water Quality

Beatrice Marinelli

Numerous definitions for indicators are mentioned in the statistical, economic, and environmental literature, though everyone agrees that an **indicator** provides a synthetic representation of reality through a value or parameter, and the information deriving from such a value is wider than the value itself and should be related to the type of user and the context in which it is situated.

Specifically, the indicators used to monitor the **qualitative state of the air** are generally those pollutants that have short-term effects on human health, such as carbon monoxide (CO), nitrogen dioxide (NO_2), sulfur dioxide (SO_2), ozone (O_3), suspended particulates (suspended particulate matter, PM_{10} or $PM_{2.5}$, depending on the dimensions), benzene (C_6H_6), and heavy metals. These indicators are based on data regarding the concentration of **atmospheric pollutants** measured at monitoring stations distributed throughout the territory; the control of these parameters is regulated by legislative decree no. 155/10, which imposes the concentration limits of the pollutants in the atmosphere.

This type of direct analysis should, however, be integrated with a type of monitoring known as biomonitoring, which uses living organisms; for the air component, the **biological indicators** most often used are epiphytic lichens, making use of the symbiotic associations between sac fungi, unicellular green algae, and cyanobacteria, which manage to accumulate appreciable levels of atmospheric contaminants (metals, radionuclides, chlorinated hydrocarbons, fluorides, and so on) that are difficult to measure in air samples. In addition, lichens, as organisms which grow very slowly, reflect with high precision the average pollution that an area was exposed to over a period of several years. The response of lichens to atmospheric pollution is quantified using the **index of atmospheric purity** developed by De Sloover and LeBlanc, which provides a quantitative estimate of the level of atmospheric pollution based on the number, frequency, and tolerance of lichen species present in the area considered.

Relative to the **quality of surface waterways**, legislative decree no.

152/99 has provided the only official national operating tool for the collection and analysis of water data. One of the chemical/microbiological indicators used for flowing waters is based on the analysis of seven descriptors known as "macrodescriptors:" dissolved oxygen (percentage saturation), biological oxygen demand, chemical oxygen demand, ammoniacal nitrogen, nitrates, total phosphorous, and fecal coliform bacteria. Each of these parameters is associated with a pressure or pollution cause, which is linked to different causes and origins and acts on the water system.

Together with the indices listed above, the international approach often uses ornithic communities as ecological indicators for ecosystem processes and **environmental monitoring** [1, 2]. Birds, in fact, are useful indicators for monitoring environmental changes and the condition of ecosystems because they often respond to the cumulative effects of environmental influences on the ecosystem [3, 4]. Found at the top of the food chain and sometimes closely depending on water resources for food, birds are notably exposed to the risks connected to **water pollution**, in addition to those inherent with the bad management of water, such as a sharp variation in water levels, for example.

References

1. Audubon Society (2004) State of the birds USA . Audubon Magazine September–October: 45–50
2. Environmental Protection Agency (2006) http://water.epa.gov/lawsregs/lawsguidance/cwa/tmdl/2006IRG_index.cfm#documents
3. Farina A, Meschini E (1985) Le comunità di uccelli come indicatori ecologici. Proceedings of the III Italian Conference of Ornithology, pp. 185-190
4. Swarth C (2003) Biological Assesment Methods for birds. Athlantic City, New Jersey

8.6.6 Evaluation Methods for Environmental Quality Assessment

Piersebastiano Ferranti

In the **environmental evaluation** field, it is normal practice to subdivide the environmental issue into items defined as elementary components or environmental issues. Such subdivision allows a simplified treatment of each environmental component, thus making possible the correct communication between the different social/disciplinary sectors.

The issue of **environmental quality** is in fact very wide and it continuously evolves in parallel to deterioration phenomena, such as **pollution**, **deforestation**, desertification, **biodiversity loss**, stratospheric ozone depletion, and climate change.

Table 8.6.6.1 Scoring system for the evaluation of building eco-sustainability

Environmental issues	Component's indicator
Environmental surroundings	**Visual-perceptive comfort** Integration in the context
External air quality	Local atmospheric pollution
Electromagnetic fields	**Electromagnetic pollution** (high/low frequency)
Acoustic exposition	**Acoustic pollution**
Ground quality	**Ground pollution**
Water quality	**Water pollution**

A common identification of the elementary components is represented by: underground and surface water; atmosphere; noise, vibrations; radiations; ground, underground; vegetation, wildlife or, generally speaking, ecosystem; and public health.

The identification is not univocal, and it is usually adapted to the specific situation examined. As an example, the subdivision into components of the external environmental quality the region of Tuscany opted for in the guidelines for sustainable constructions, is reported in **Table 8.6.6.1**. Such components identify a scoring systems for the evaluation of **building eco-sustainability**.

It is clear how the issue of the quality of the environmental components is faced by means of the measure of the reciprocal concept, i.e., pollution. For instance, air quality is related to atmospheric pollution, which is intended to be "*the air quality status consequent to the introduction of any substance in such measure and conditions to alter air healthiness and to constitute direct/indirect prejudice for human health, as well as damage for public/private goods*".[15]

Taking into account the example of the atmosphere–air quality component, the concept of indicator to achieve the complete definition of an evaluation method for environmental quality has to be introduced at this point.

The use of indicators by human beings is a solid tool of knowledge and interpretation. The indicators are fundamental in organizing the information for the methods finalized to environmental evaluation (EIA-Environmental Impact Assessment; SEA - Strategic Environmental Assessment) or environmental reporting. Indicator means an empirically observable, calculable, or monitorable characteristic (or pool of character-

[15] APAT (2006) Qualità dell'ambiente urbano III rapporto. APAT Edizione (APAT: Italian environmental protection agency).

istics) through which it is possible to describe a certain phenomenon. The indicators must be representative of the current as well as the final environmental status, so that a simple comparison operation makes it possible to identify the variation. Both indicators and environmental components are selected via a scoping phase, depending on the issue under examination. In this phase, consultative activities with the local communities, aimed at integrating the knowledge of the evaluating experts become important.

Returning to the atmosphere–air quality component, about 3,000 air-contaminating substances have been classified, mostly deriving from human activities. The principal pollutants include carbon monoxide, sulfur and nitrogen oxides, benzene, ozone, and particulate matter.

Within the frame of the **core environmental indicators** for urban air quality description, the Organisation for Economic Co-operation and Development (OECD) uses air pollutant concentrations as a status indicator in relation to population exposure to such pollutants. The OECD uses the so-called pressure-status-response (PSR) model to describe a country's advance as to its environment. A frequent model for the elaboration of the VAS environmental reports is the Driving forces-Pressures-States-Impacts-Responses (DPSIR) model, an evolution of the PSR. It allows one to correlate the information describing the status and the modifications of a certain environmental context, by following a logical scheme. The DPSIR model is not "rigid," i.e., a component may have different roles (determinant, impact, and so on) depending on logical scheme construction.

A further concept to introduce for the definition of an environmental quality evaluation method is that of the threshold limit value. In the previously mentioned example regarding the atmosphere–air quality component, the threshold limit value is a limit value for the concentration of the pollutant in the atmosphere. The threshold limit values are typically identified by international research institutes. Then, each country adopts them through the legislative process.

The use of the DPSIR model therefore consists of:

1. Analyzing a "determinant" (e.g., car traffic).
2. Evaluating "pressures" throughout the assessment/direct measuring of the consequent emissions (e.g., NO2).
3. Analyzing the "status" through the direct measurement of the pollutant concentration, and subsequently evaluate current air quality.
4. Evaluating the "impact" of a potential car traffic variation through an assessment of the "status" variation (i.e., variation in the pollutant concentration), comparing it with the threshold limit values identified by scientific research and by the legislative process.
5. Hypothesizing any "responses" (mitigation and oriented measures), if needed.

8.7 A Case Study of Implementation: The Marche Regional Ecological Network[16]

The case study of the Marche Regional Ecological Network (Rete Ecologica delle Marche, REM from hereon) represents an example of area of examination titled "Distinctive and Pleasant" in chapter 8.3. It closely examines the relationships between the city and the immediate context trough the environmental infrastructure. The project objectives are:

1. Create a strategy and measurements to face the problem of protection and evaluation of the **regional environmental heritage** as a whole.
2. Define an area of intervention that concerns the entire regional territory (not just nodes and corridors), defining the forms of contact with landscapes in the Marche region.
3. Manage the regional environmental system, governing the functions of geographical and typological unity.
4. Guarantee a system of connections that ensures the functioning of the REM for the complex of regional **biodiversity** by using both focus species and species of conservation interest.
5. Establish modes of interaction between the REM and regional planning and programming tools.

The project is developed on three different, yet complementary levels of examination:

- The **inter-regional scale**, to highlight the role of the REM in the **National Ecological Network (REN)**.
- The regional scale, to portray the Marche Region's ecological system and its necessary connections to the landscape plan and the regional territorial plan for placement within the framework.
- The **local scale**, to handle specific themes and propose implementation projects, even with symbolic worth, in close connection with municipal urban planning tools and territorial plans for provincial coordination.

Formation of the REM is articulated in three phases (recognition framework, interpretive framework, and project framework) and is developed through two complementary paths:

- The first is an approach that integrates phytosociological information with faunal and anthropic aspects to reach the definition of **ecological-functional unity** that is the basic element of the perspective part of the REM. A characterization of the entire territorial matrix as a function of the ecological network can thus be obtained.

[16] The working group who drafted the REM project is composed of officials from the Marche region, professionals from the University of Camerino (UNICAM) spin-off "Terre.it," and researchers from the following universities: Ancona, Camerino, Macerata, Torino, and Urbino (general coordinator: Claudio Zabaglia; scientific coordinator: Massimo Sargolini; technical coordinator: Paolo Perna).

- The second takes its movements from the **nodes** of the REM. They derive from the integration of the system of areas of environmental prestige with other areas of particular biodiversity importance identified within the REM.

The cognitive frameworks are organized into two large systems: biological and anthropic. For the first, the following maps have been produced: unit of vegetation landscape (Fig. 8.7.1); potential vegetation (Fig. 8.7.2), naturality and geobotanical worth (Fig. 8.7.3); habitat of community interest (Fig. 8.7.4); and average cenotic faunal index. For the second, the following maps have been produced: settlement organization (Fig. 8.7.5); landscape fragmentation: index of infrastructure fragmentation (Fig. 8.7.6); landscape fragmentation: index of urban fragmentation (Fig. 8.7.7); potential environmental fragmentation: sensitivity to settlement development (Fig. 8.7.8).

The **botanical analysis**, through the interpretation of the phytosociological map of vegetation, forms the basis for the ecosystem reading on which the REM is based. The faunal investigations have been realized from two different points of view (which converge in the project phase): animal communities and target species. The investigations regarding the anthropic organization have highlighted human settlement forms and modes in relationship with the socio-economic systems of reference, also highlighting elements and contexts of potential worth or criticism. In addition, the analysis of the ongoing planning and programming is focused on the following items: territory and landscape, infrastructure organization, and sectoral planning.

The **interpretational framework**, by integrating different cognitive frameworks, identifies the characterization of the environmental system through:

1. Reading the narrative of structures and the relationships that form the regional ecological fabric. This is a **transdisciplinary vision** within the regional territory, with the aim of highlighting the mechanisms and processes at the basis of the functioning of the biological system in the Marche Region. It is noteworthy that while the main **ecosystem units** (Fig. 8.7.9) are located near the Apennine ridges, where the large transregional connection systems (Fig. 8.7.10) are sited, the functional ecological units are of interest to the entire territory, including the large settlement areas on the Adriatic coast (Fig. 8.7.11).
2. Reading the **nodes/corridors**, which aims to highlight portions of the territory with special functions with respect to species and groups of selected species within the ecological fabric. This has led to the definition and characterization of nodes with respect to environmental systems. It is noteworthy that the entire territory is interested in the presence of nodes that are important for **bioconnections**: those related to the grassland (Fig. 8.7.12) and rupestral environments (Fig. 8.7.13) prevalent in the mountainous area; those related to forest environments in the foothill zone (Fig. 8.7.14);

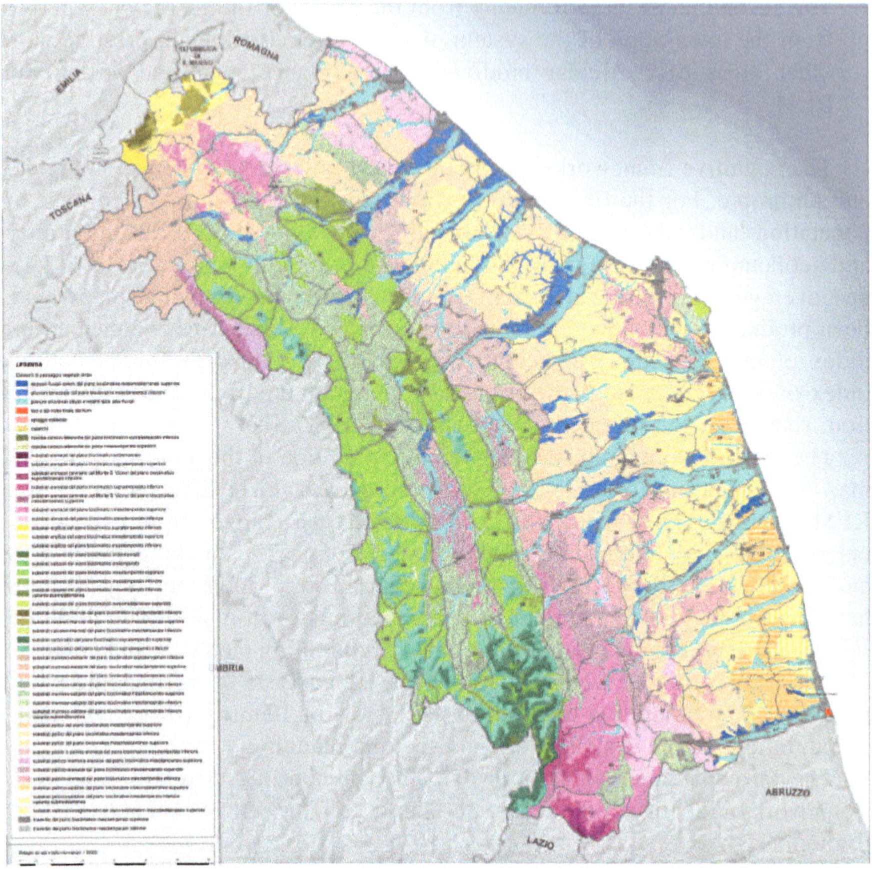

Fig. 8.7.1 REM (Marche Environmental Network). Vegetation Landscape Units (ULV). ULVs were defined using a complex from the vegetation map (defined by different relationships between geology and bio-climate units), complemented by relationships with settlement-covered surfaces. *Blue* represents ULVs based on fluvial and alluvial deposits; *violet* represents those based on sandy substrates; *yellow* for those based on clay substrates; *green* represents those based on calcareous substrates; *gray* represents those based on travertine

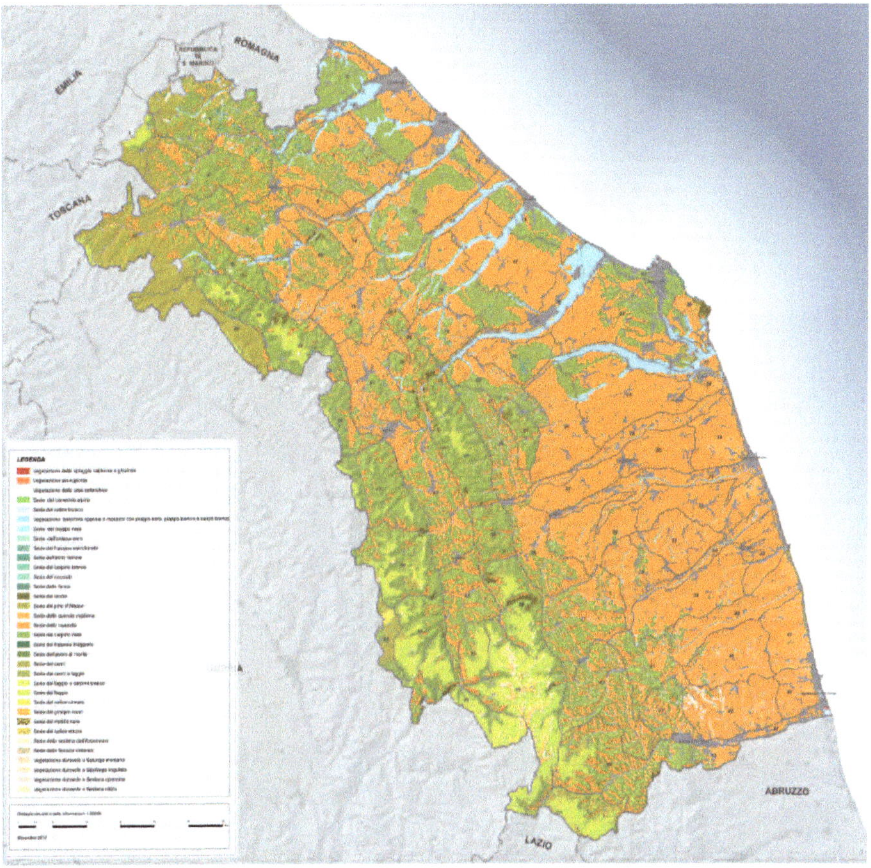

Fig. 8.7.2 REM (Marche Environmental Network). Potential vegetation. This allows for the evaluation of vegetation formation in each part of the territory, which would expand the system if the disturbance ceased. This is very useful, because it allows the existing ecological differences in different parts of the region to emerge as well as indicating the most appropriate actions with respect to the natural potential for the realization of interventions aimed at consolidating the environmental systems

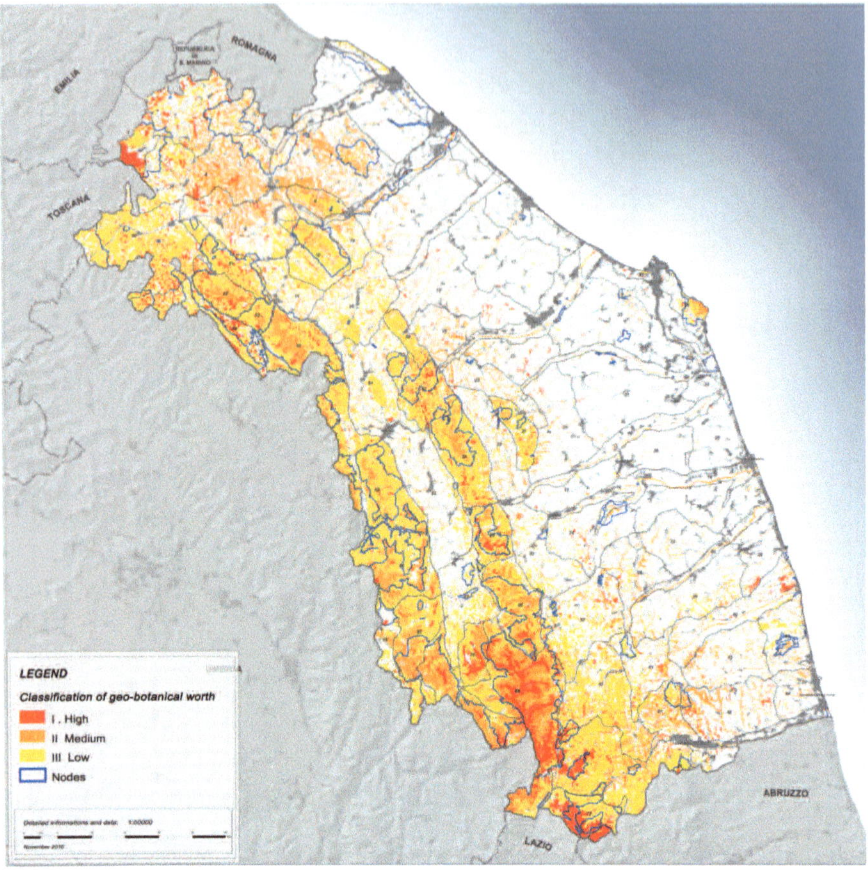

Fig. 8.7.3 REM (Marche Environmental Network). Geobotanical worth. The map of geo-botanical worth represents, together with the maps derived from it (maps of landscape structure, naturalness, and landscape trajectories), a valid tool for evaluating the level of environmental quality and vulnerability in the study area, and it allows the identification of the most important areas for the conservation of phytosociological biodiversity, which would contribute to the construction of the REM node system. Close to *red* are the classes of greatest worth; close to *yellow* are lower classes. The nodes are outlined in *blue*

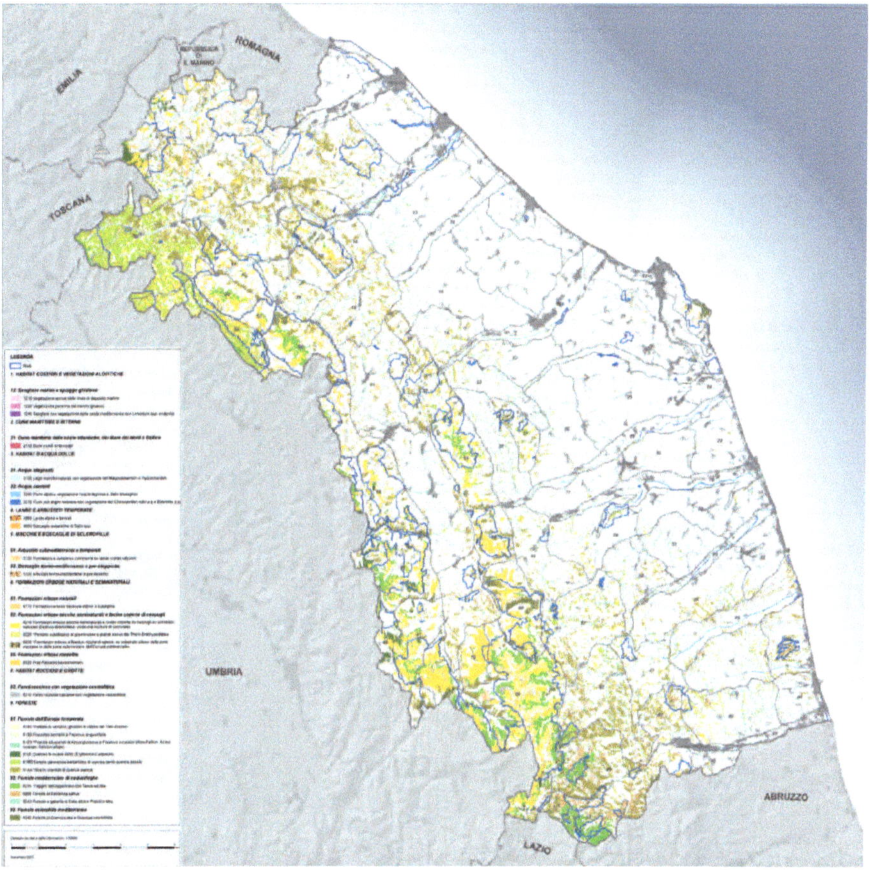

Fig. 8.7.4 REM (Marche Environmental Network). Habitat of community interest: marine cliffs and pebbly sands (*in the violet scale*); maritime dunes on the Adriatic coast (*red*); freshwater habitats (*blue*); heaths and temperate shrubs, brush, and undergrowth (*beige*); grassy formations (*yellow*); rocky walls (*gray*); forests (*green*)

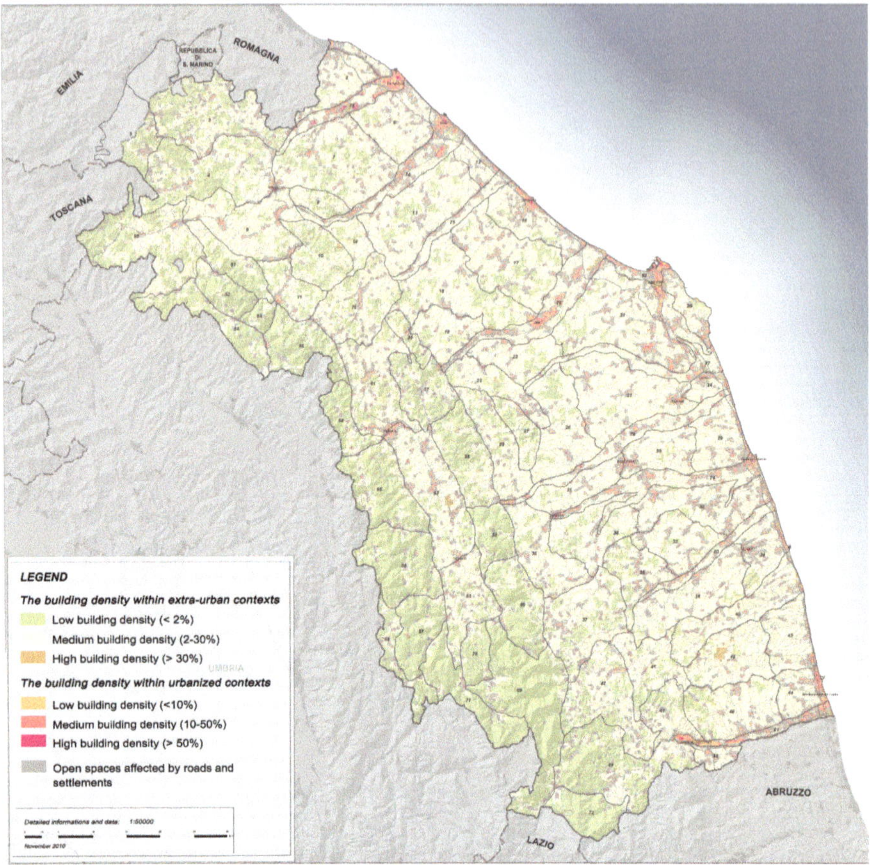

Fig. 8.7.5 REM (Marche Environmental Network). Settlement organization. Zones of different building density are identified: in suburban (*brown*) and urban areas (*red*). Areas of medium density match the prominent areas. These are zones in which the building history, more or less diffuse and free of centralizing opportunities, merges with cultivation and floor management systems. In such a context, one is left identifying areas that brush the edges of the major urban centers, whether they are plains, knolls, or hillsides. The areas of highest density, which concentrate social dynamics, are noted in provincial capitals. In satellites connected to them, a progressive increase in importance from the functional (dormitory city and industrial poles under transformation) and logistical point of view is noted

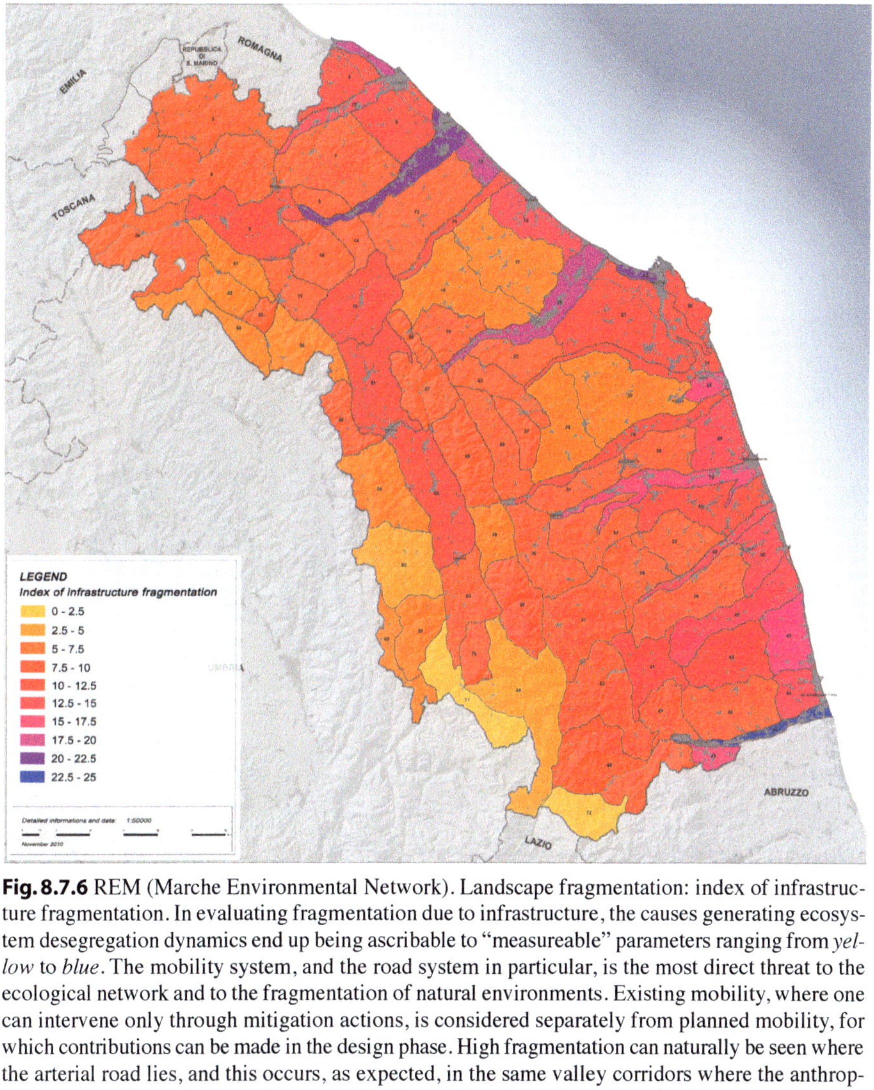

Fig. 8.7.6 REM (Marche Environmental Network). Landscape fragmentation: index of infrastructure fragmentation. In evaluating fragmentation due to infrastructure, the causes generating ecosystem desegregation dynamics end up being ascribable to "measureable" parameters ranging from *yellow* to *blue*. The mobility system, and the road system in particular, is the most direct threat to the ecological network and to the fragmentation of natural environments. Existing mobility, where one can intervene only through mitigation actions, is considered separately from planned mobility, for which contributions can be made in the design phase. High fragmentation can naturally be seen where the arterial road lies, and this occurs, as expected, in the same valley corridors where the anthropic system is principally seen. However, situations of heightened value can be detected that closely correspond to the edges of built areas in some main urban centers

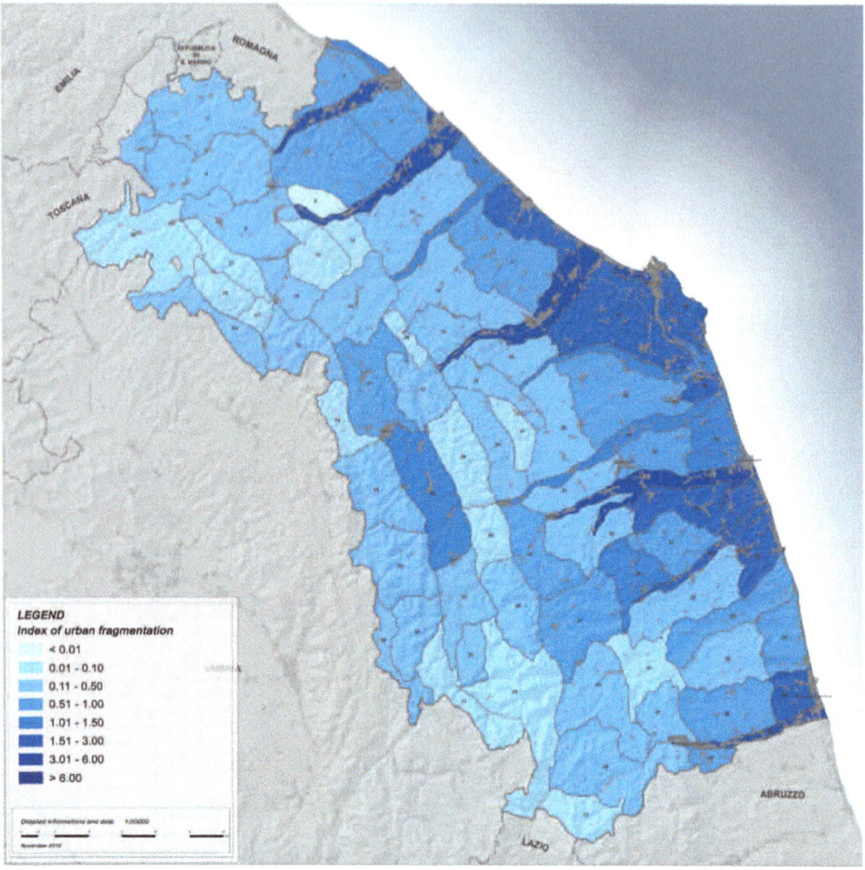

Fig. 8.7.7 REM (Marche Environmental Network). Landscape fragmentation: index of urban fragmentation. Numerical values are related to the surface area and length of characteristics constituting settlement organization. The linear urbanization fragmentation index (UFI) is measured on the basis of environmental units in which the regional territory is divided. The success of the investigation exposes a pronounced fragmentation phenomenon corresponding to the main access channels that penetrate the interior from the coast, or rather, in the main valley areas (tracing out the typical comb structure of the Marche Region). The phenomenon, which is considerable close to the coast (*dark blue*), weakens in moving from the plain or low hills to the high hill and foothill zone (*light blue*). However, critical situations remain, even inland, such as what happens in the case of the syncline that reaches from Camerino to Pergola

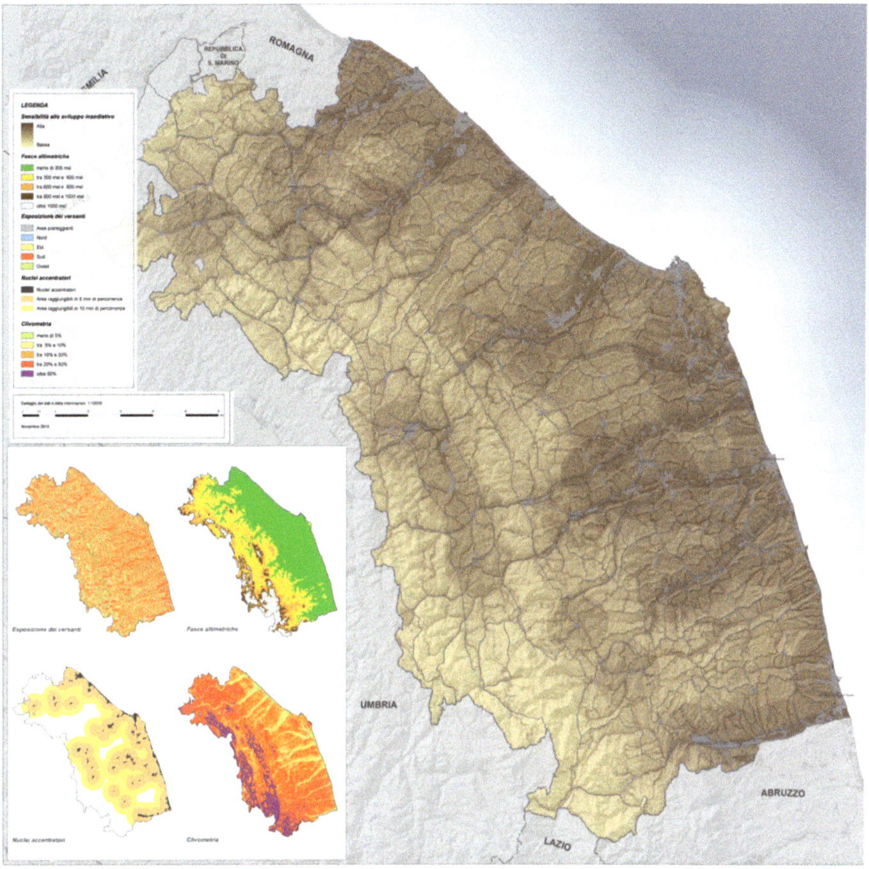

Fig. 8.7.8 REM (Marche Environmental Network). Potential environmental fragmentation: sensitivity to settlement development. We intend to evaluate the dynamics of the sprawl phenomenon, identifying measures to quantify its expansion. The sensitivity of the territory to undergoing processes of land consumption has been examined, taking as a reference morphological parameters such as acclivity, the exposure of hillsides, and altimetry[17]. The weighted sum of the indices of each category serves to define the associated value of the overall index of sensitivity to urbanization. Dark brown denotes high sensitivity to urbanization. The settlement and infrastructure history have, in many cases, already shown the autonomy of ecological networks, halting its push

[17] Battisti C, Romano B (2007) Frammentazione e connettività. Città Studi Edizioni. Novara.

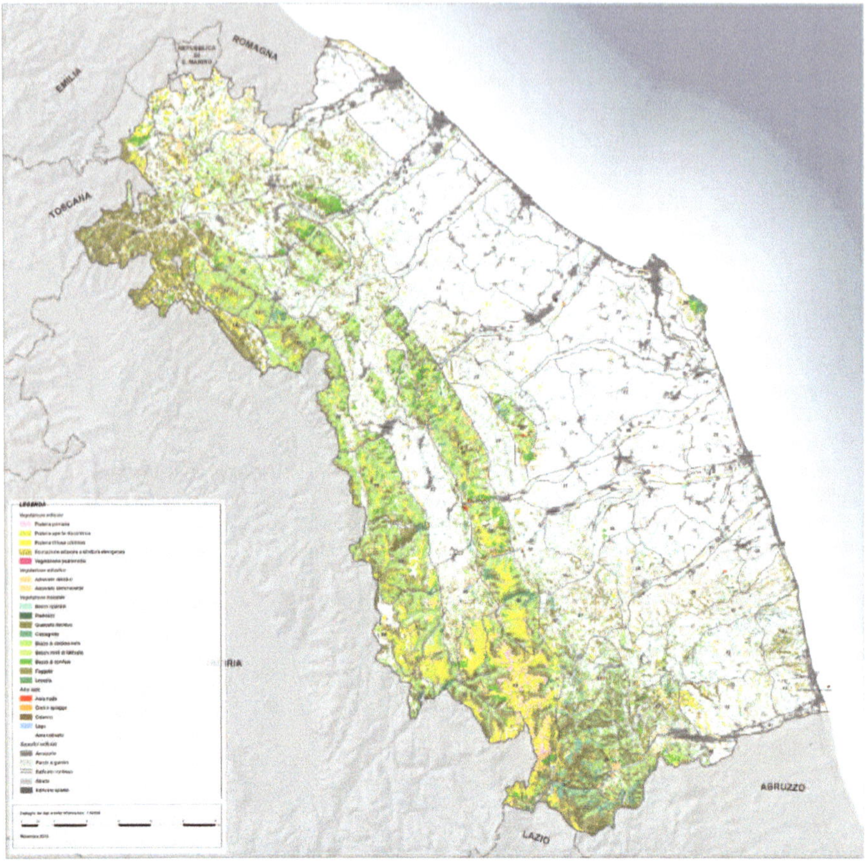

Fig. 8.7.9 REM (Marche Environmental Network). Ecosystem units. The need to define management objectives, strategies, and measures that are coherent and applicable has necessitated a synthetic classification of the elements constituting the regional ecological system, which, starting from phytosociological analysis, considers faunal information and the conditions of settlement and infrastructure organization, with particular attention to urban and agricultural areas. In this way, the map of ecosystem units covers the area, thereby giving an ecological meaning to each individual part of the territory, even in the most settled areas. Each of these undertakes an essential role for bioconnectivity that, in turn, becomes a supporting structure for urban and peri-urban areas

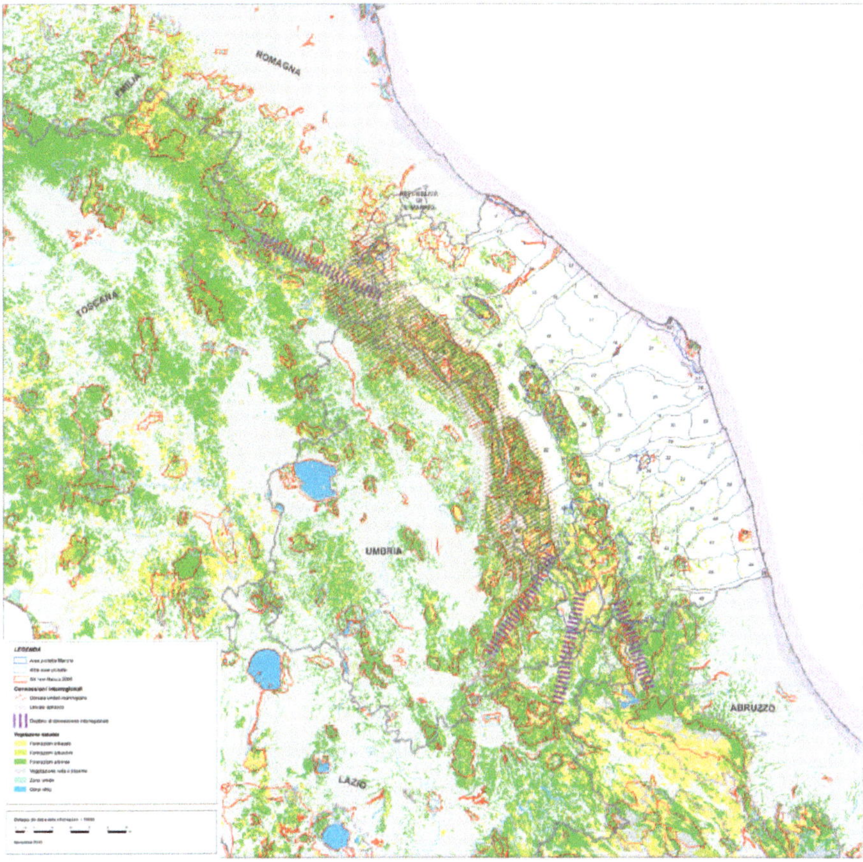

Fig. 8.7.10 REM (Marche Environmental Network). Systems of transregional access. The large connection systems of transregional interest constitute continuous natural areas that, on the one hand, are connected to the Appenine backbone and on the other, are more or less spread throughout the hill territory until they reach the coast. Modeled on the structure of mountainous systems, they are therefore determined by two long swaths, one along the Umbrian/Marche Appenine backbone, and the other along the Marche backbone, which come together near the Sibillini mountains massif. In general, they are quite extensive in the high-hill areas and gradually decline near the coast, where they are limited to the riparian zone

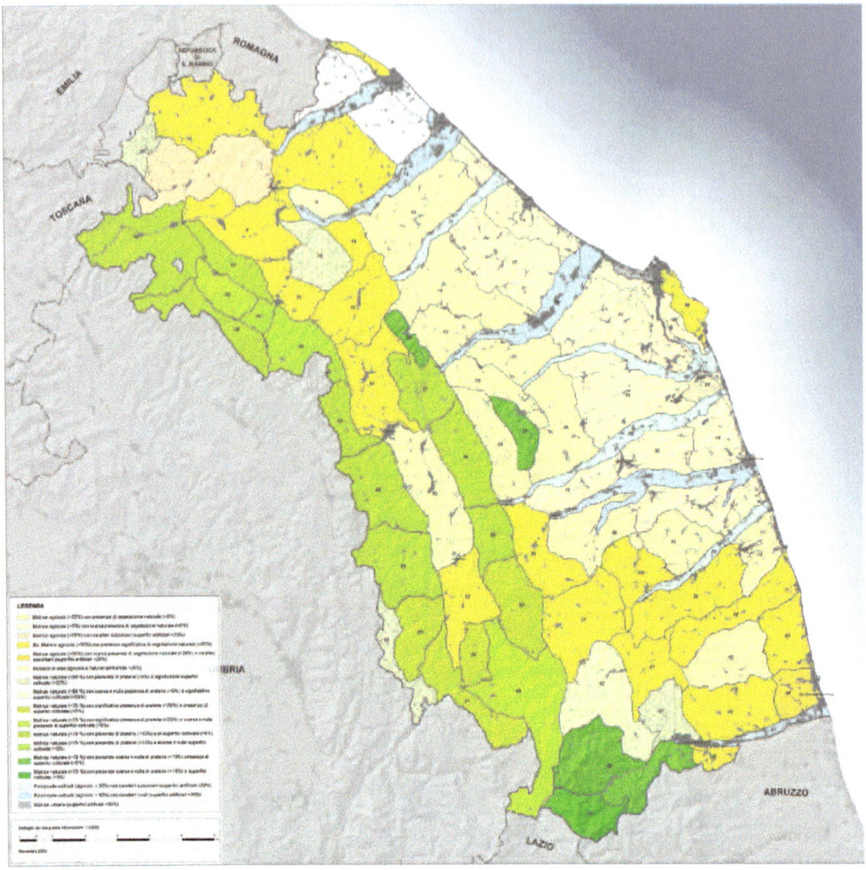

Fig. 8.7.11 REM (Marche Environmental Network). Ecological Functional Units (UEF). UEFs usually integrate information about vegetation, faunal, and anthropic aspects, characterizing the territory in its different structural and functional frameworks, highlighting how natural elements and anthropic activities are related, giving rise to the diversity of landscapes typical of the Marche Region. Through this approach, the REM becomes a special interlocutor for landscape planning tools and, in particular, for the new Regional Environmental Landscape Plan that the region is preparing. In *yellow* are the UEFs on the agricultural matrix; those on the natural matrix are in *green*; those on urban and peri-urban matrices range from *blue* to *gray*

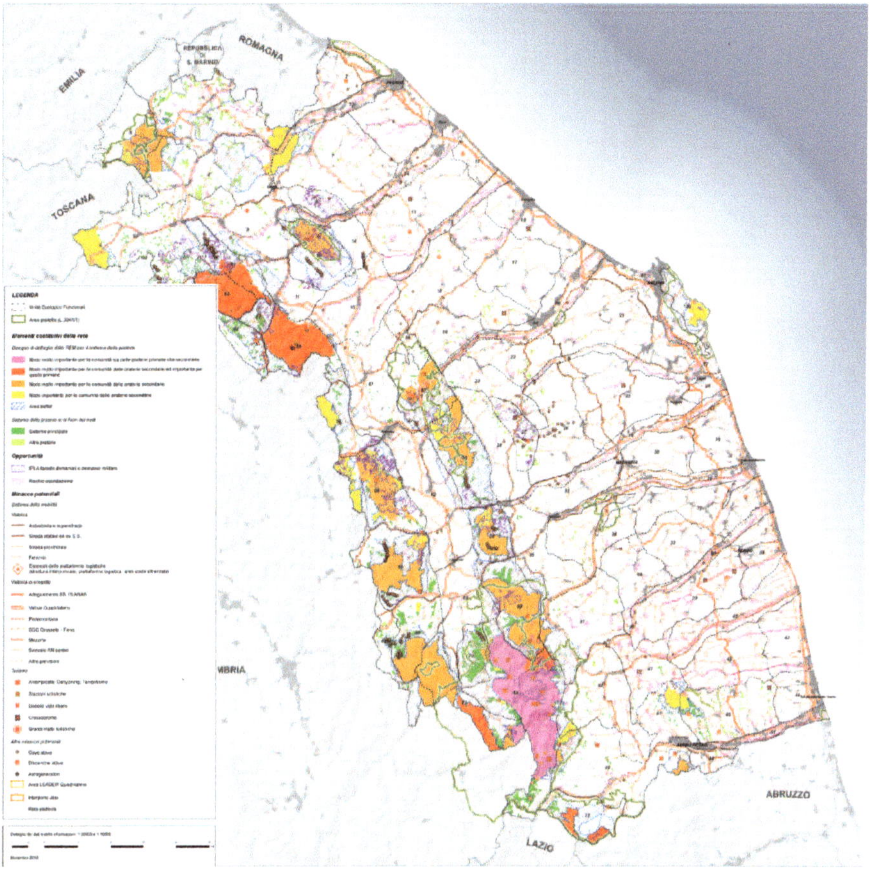

Fig. 8.7.12 REM (Marche Environmental Network). Detailed drawing of the REM for the environmental system of grasslands. Grassland formations are one of the most important elements for biodiversity conservation in the Marche Region. Grasslands are concentrated primarily in high mountain areas and because of this, they are presented as "islands" that are more or less separated from one another. Connection areas are the areas external to the nodes, whose state of conservation affects the possibility (or not) of populations to expand, thereby allowing both the survival within nodes and the flux of individuals between them. The grassland system does not lead to significant contact with the urban and peri-urban settlement structure. The importance of nodes for grassland communities is shown from *yellow* to *violet*. The grassland system outside the nodes is shown in *green*

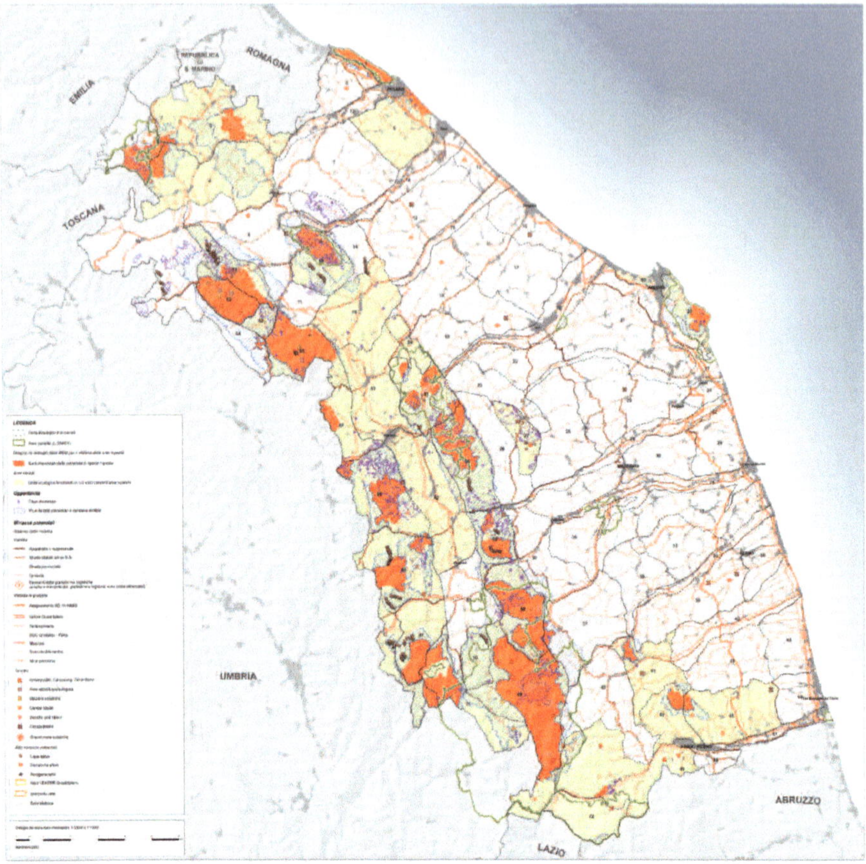

Fig. 8.7.13 REM (Marche Environmental Network). Detailed drawing of the REM for the environmental system of rupestral areas. The design of the REM for these environments was defined, highlighting the functional ecological units within which rocky areas ideal for the settlement of target species are present. The identification of these sites was postponed to subsequent phases, for a closer examination of the design of the networks by competent territorial subjects. The nodes of interest for the presence of rock-inhabiting species are shown in *red*. Functional ecological units that include rupiculine areas are shown in *light green*. Monte Conero and the cliff in the Fano/Pesaro zone, where the interaction between settlement structures and UEFs is very intense and determines the landscape character of the area, are particularly interesting for landscape planning ends

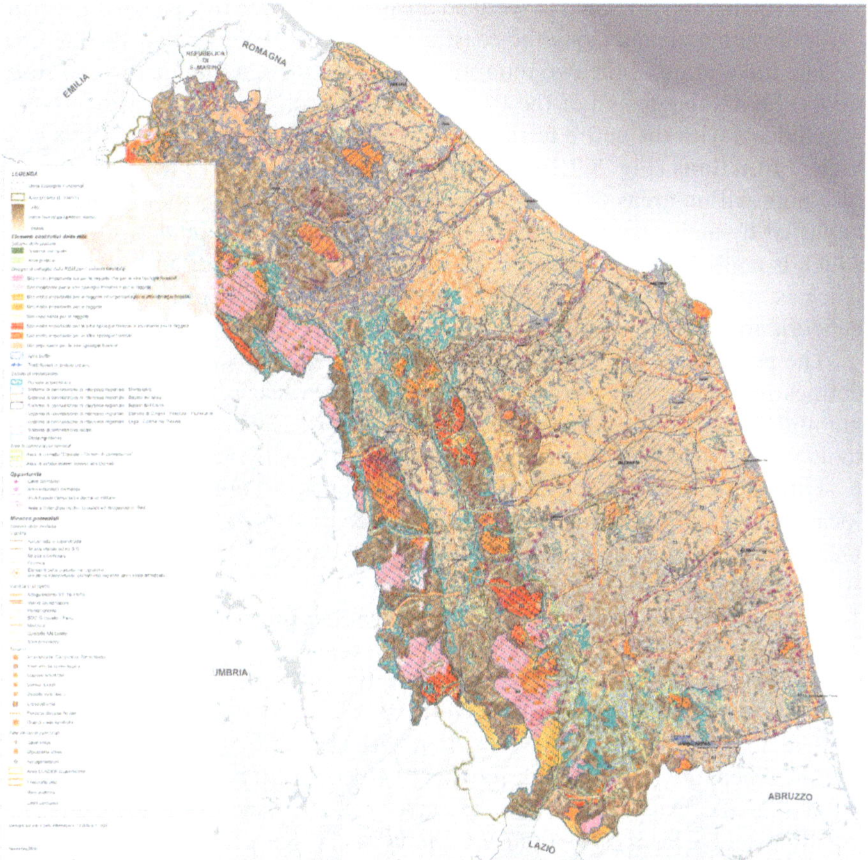

Fig. 8.7.14 REM (Marche Environmental Network). Detailed drawing of the REM for the environmental system of forests. The forest system is certainly the most interesting due to the number of species present and for its diffusion throughout the region. Forests, or their remaining fragments, spread throughout the hill areas and through areas significantly modified by anthropic processes, often reaching to the sea, and significantly contribute to the biodiversity of agrarian landscapes. This system, which is configured as a supporting element of the REM, intercepts and contaminates many urban and peri-urban spaces, both in hill settlements and settlements along valleys and the coast. The different grades of importance of forest nodes are shown from *orange* to *violet*. The different colored edges delineate various systems of forest connection

and finally, those related to wetland areas (Fig. 8.7.15) that are mostly found in the hill zone and along the coast, and therefore in the most settled areas.

The integration of the two interpretations (overall and limited to nodes/corridors) favors the design of the REM, which is configured as a base narrative which deals with the entire territory, and is characterized according to its **ecological functions** (Fig. 8.7.16). The elements constituting the network are diffuse both within areas of great naturalness and within those more extensively modified by anthropic processes. For example, near the capital of the region, which is considered a coexistence of metropolitan and important natural areas (such as the Conero Regional Park), almost all of the elements constituting the network are in close proximity to one another (Fig. 8.7.17).

The perspective framework is arranged in three distinct parts: the first part defines the measures to be adopted for the individual resources; the second highlights urgencies and priorities at the territorial level; the third identifies tools for activating the monitoring system at different levels.

Finally, to meet the objective of favoring the activation of the REM at the local scale, 25 project investigations have been developed, which carry particular importance for the significantly urbanized areas. These are divided into five emerging themes and contexts:

1. The coastal city and its residual **environmental relationships** with the hills.
2. The settled valley floor, river connectivity, and green backbones.
3. The agrarian landscapes and the diffuse connectivity of inland areas.
4. The dilated Apennines: the transition between the ridges and the foothills.
5. The Apennine ridges and the connection between protected territories.

Within the case studies coming under theme 1, we have identified the "Colline del Fermano" as a typical example. This is an area of strong commercial productivity, morphologically distinct from the settled valley floor. Whether on the crest of the hills or in the valley, a progressive joining together of built areas is seen. The rural area of the remaining part of the territory coexists with elements constituting the regional ecological network (Fig. 8.7.18). Strategic orientation primarily deals with interventions to reinforce the connections between the components of the **river network**. Particular attention is placed on the **residual internal areas** to counteract the risk of a reduction in biopermeability. Vegetation formation is the object of particular attention, above all when continuity could be lost by intersecting various infrastructures (Fig. 8.7.19).

Within the case studies coming under theme 2, we have identified the floor of the Tronto Valley as an example, which is characterized by the tight procession of natural and man-made components. The river corridor marks the southern limit of building expansion, which in a longitudinal sense, tends toward the progressive joining together of its parts, above all near the arteries of the infrastructure. The vegetation formation of the hydrographical network and the division of cultivation into farms, keeps the relationship between the river and the rural and urban fabric alive (Fig. 8.7.20). From the strategic point of

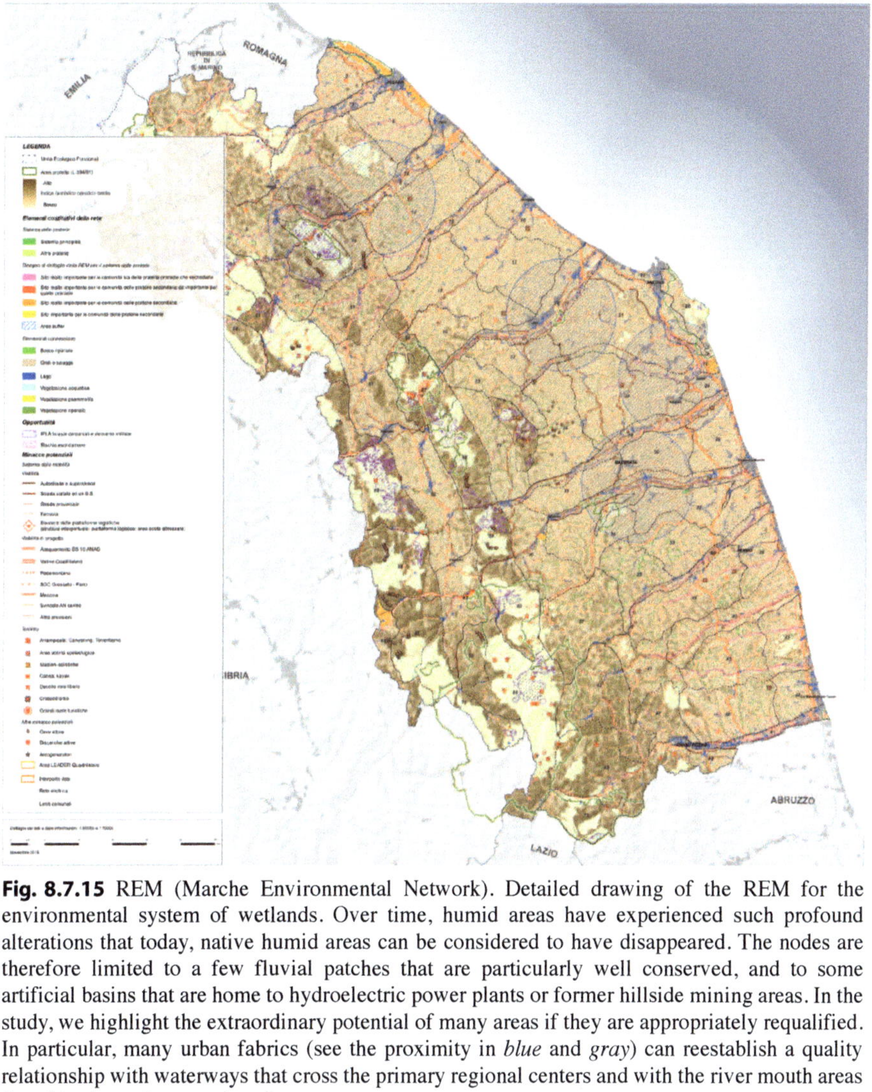

Fig. 8.7.15 REM (Marche Environmental Network). Detailed drawing of the REM for the environmental system of wetlands. Over time, humid areas have experienced such profound alterations that today, native humid areas can be considered to have disappeared. The nodes are therefore limited to a few fluvial patches that are particularly well conserved, and to some artificial basins that are home to hydroelectric power plants or former hillside mining areas. In the study, we highlight the extraordinary potential of many areas if they are appropriately requalified. In particular, many urban fabrics (see the proximity in *blue* and *gray*) can reestablish a quality relationship with waterways that cross the primary regional centers and with the river mouth areas that skirt the main coastal settlements

view, the **residual agricultural areas** or those marked by natural values, which are found compressed between urban residential areas and productive/commercial areas, undertake a role to preserve the relationship between the river corridor and the fabric modified by humans. These same areas are the object of interventions for the **mitigation of impacts** generated by the extension and joining together of the built areas (Fig. 8.7.21).

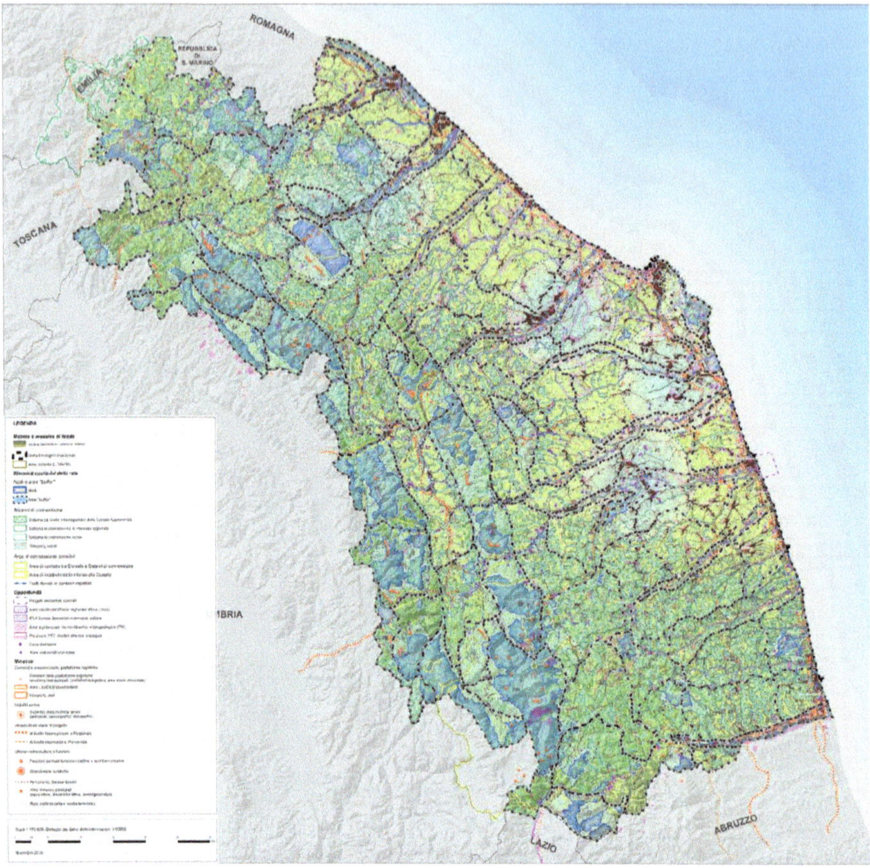

Fig. 8.7.16 REM (Marche Environmental Network). General design and layout of the network. The following have been highlighted: valley floor matrix, elements constituting the network (nodes, buffer areas, connection systems, sensitive connection areas), and opportunities and threats, with particular attention placed on the planned road network infrastructure, logistical platforms, centrality and tourist facilities, and industrial and commercial expansion

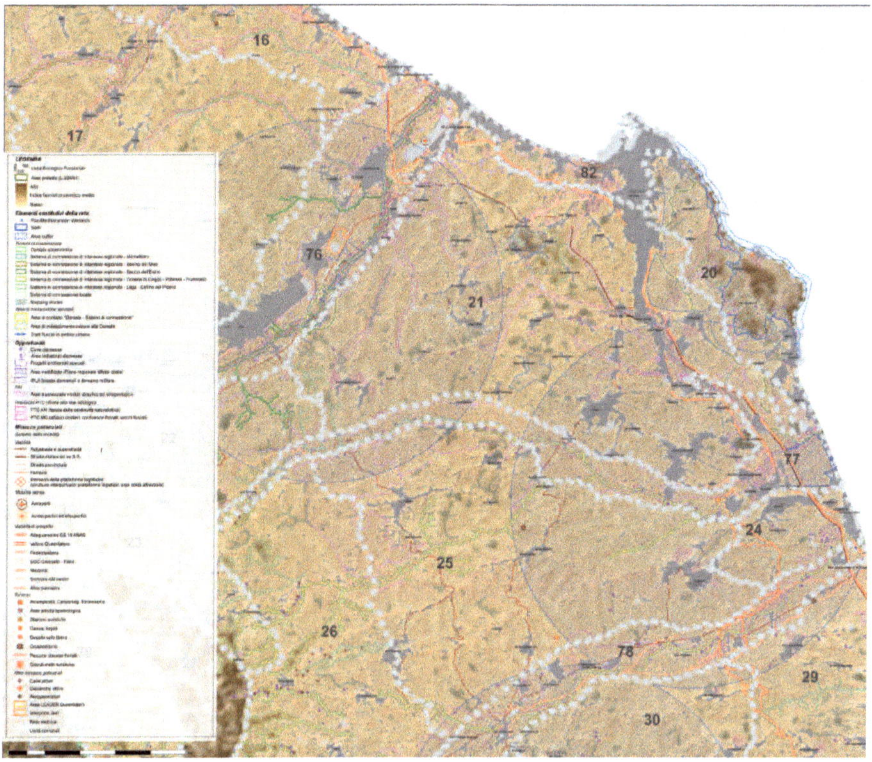

Fig. 8.7.17 REM (Marche Environmental Network). Detailed design of the network of Functional Ecological Units. Some areas of closer examination: 20 - Monte Conero, 21 - Hills of Santa Maria Nuova and Osimo, 22 - Hills of Cupramontana and Santa Maria Nuova, 77 - Fondovalle Musone, 82 - Ancona

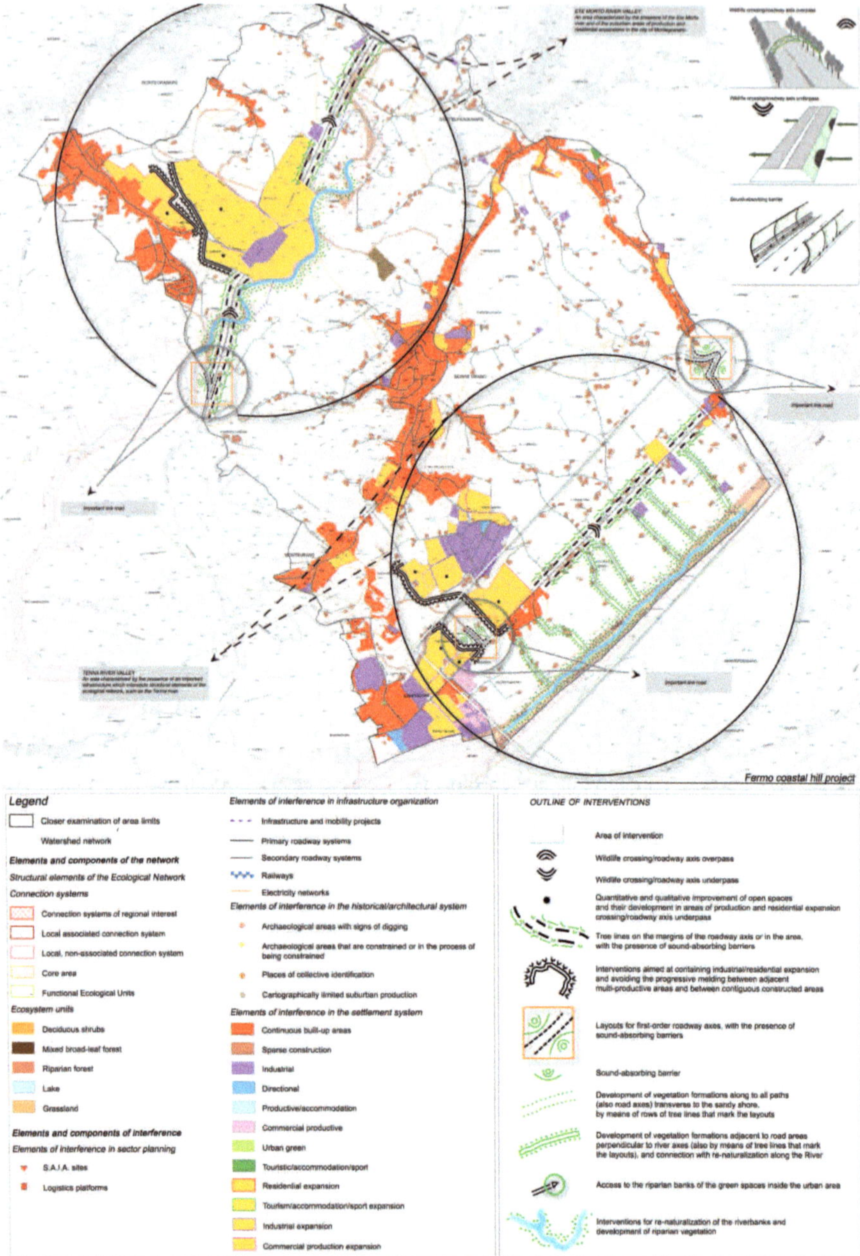

Fig. 8.7.18 REM (Marche Environmental Network). Project area: *Fermo costal hill* - project objectives

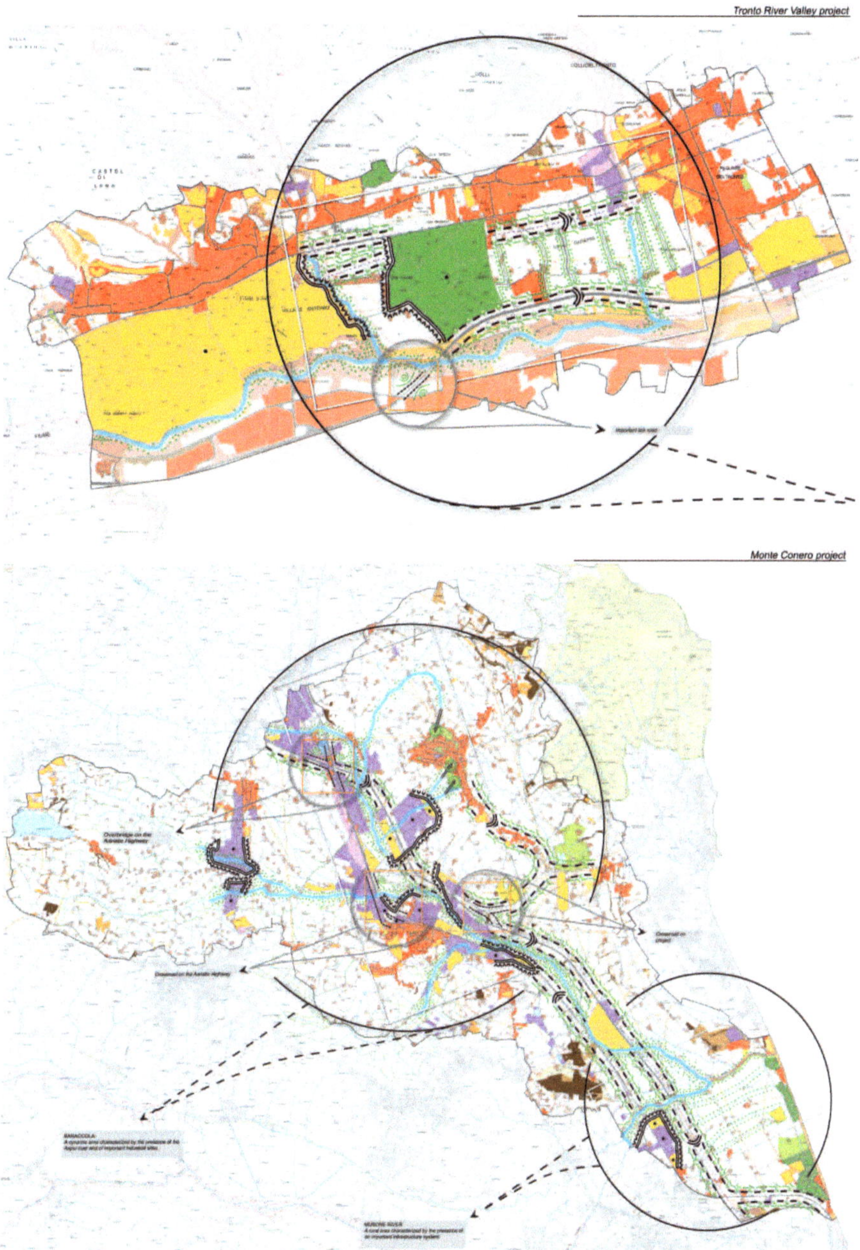

Fig. 8.7.19 REM (Marche Environmental Network). Project areas: *Tronto River Valley* and *Monte Conero* - project objectives

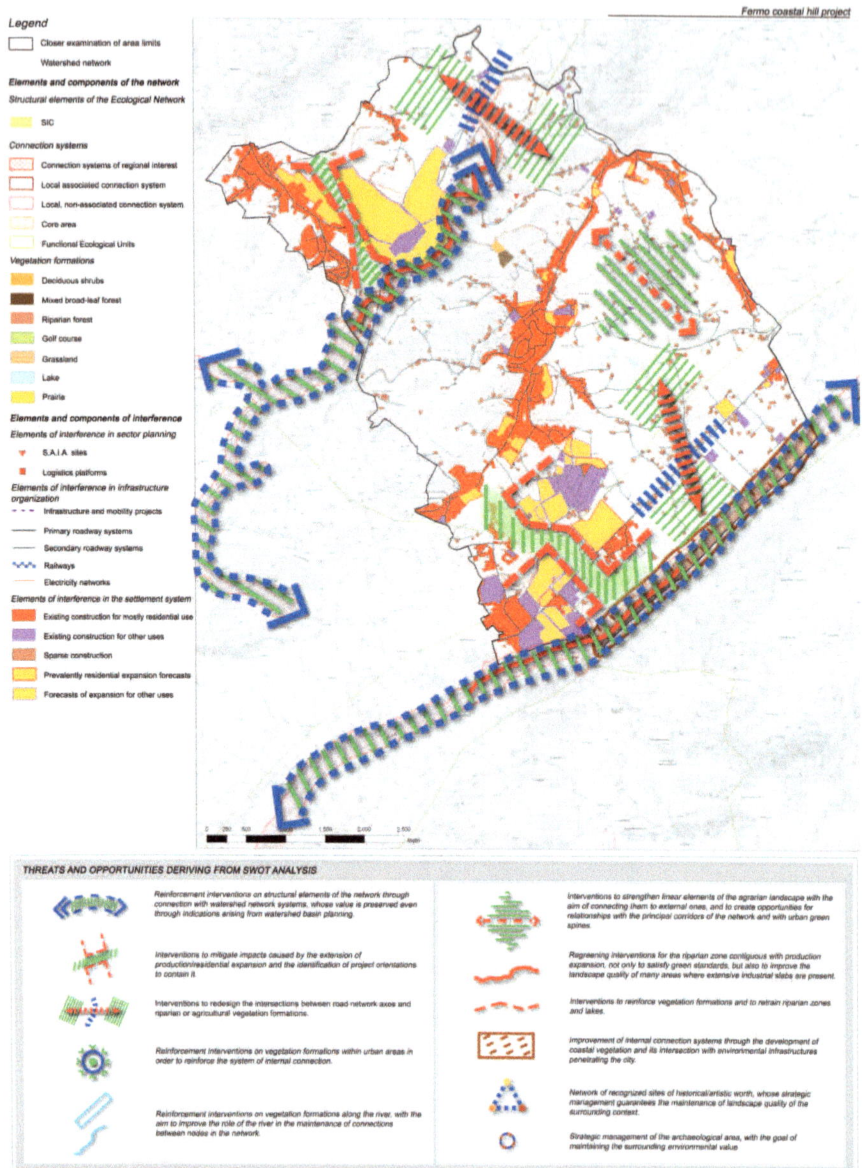

Fig. 8.7.20 REM (Marche Environmental Network). Project area: *Fermo costal hill* - threats and opportunities

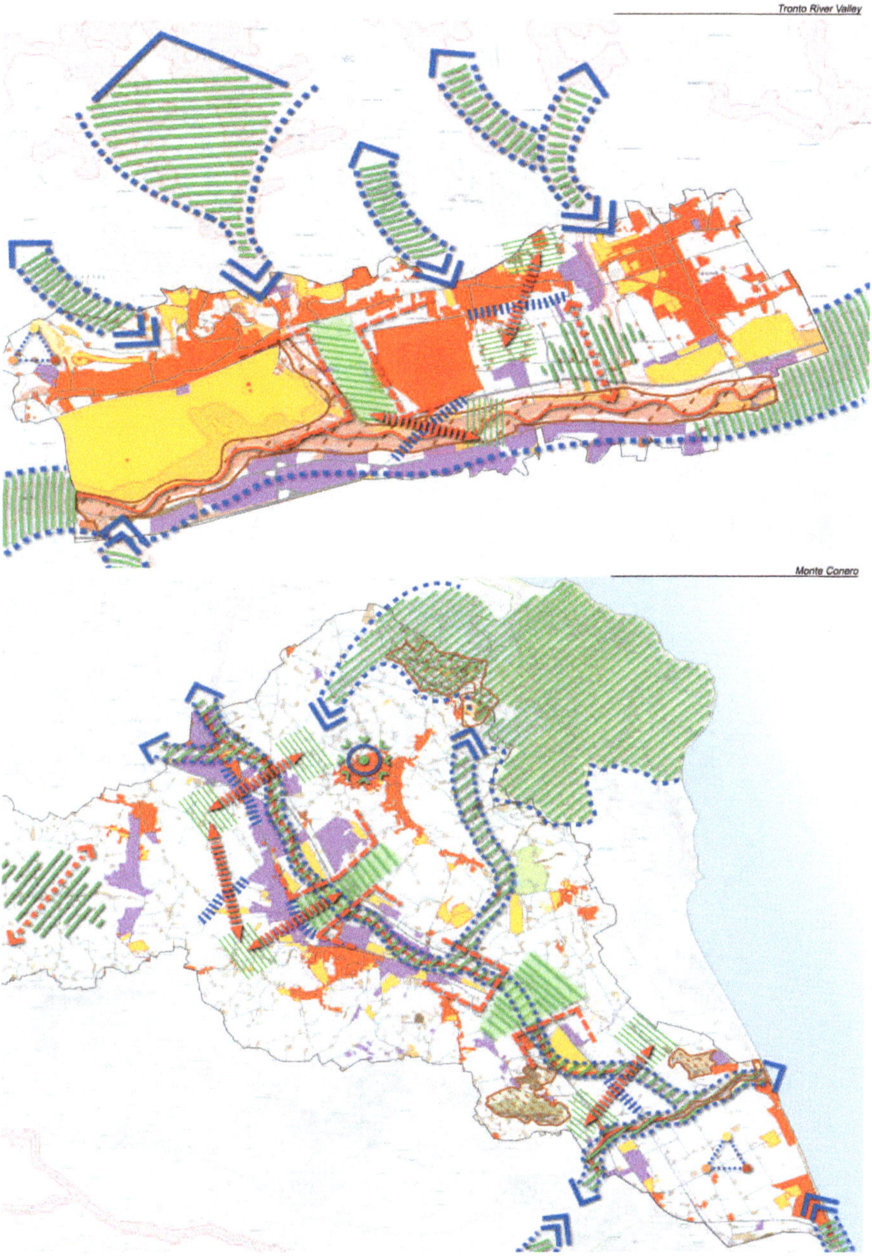

Fig. 8.7.21 REM (Marche Environmental Network). Project areas: *Tronto River Valley* and *Monte Conero* - threats and opportunities

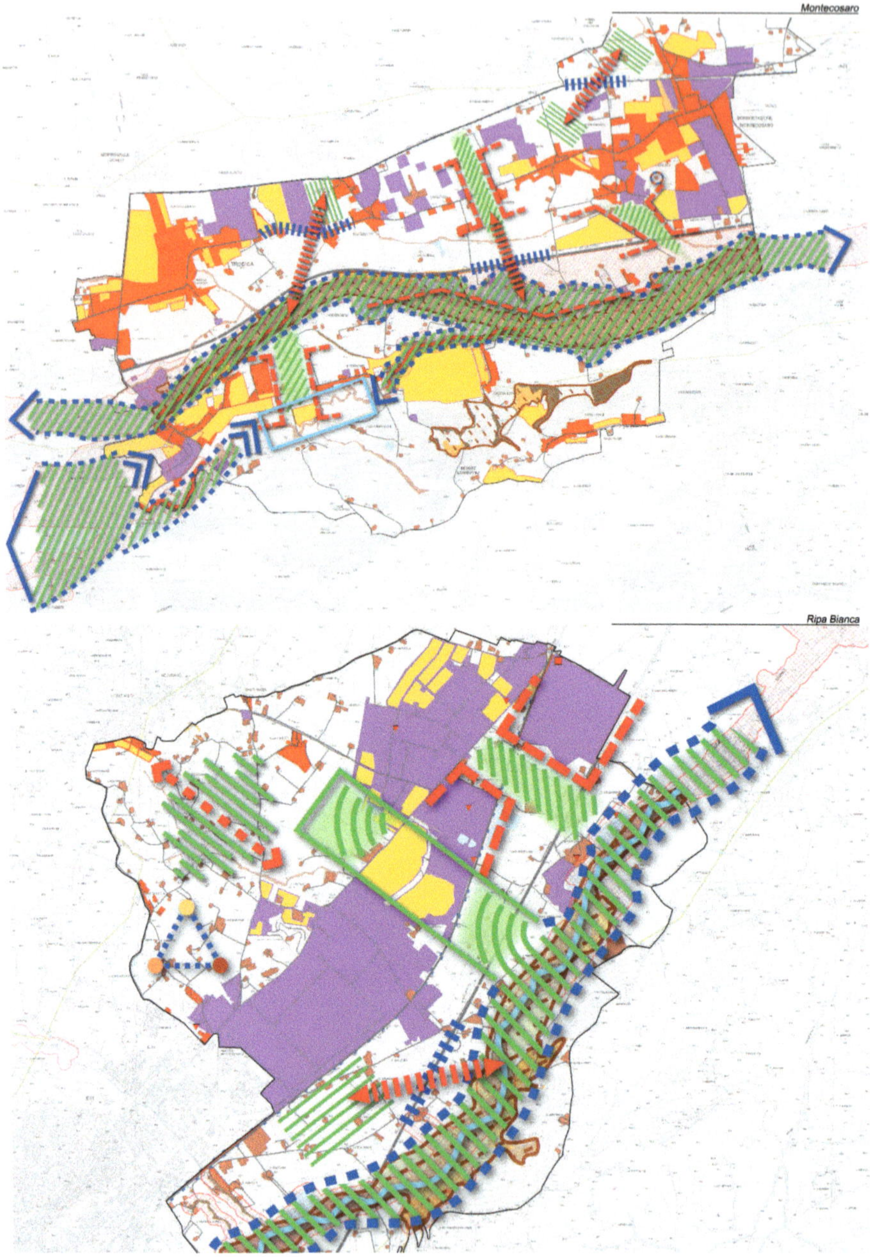

Fig. 8.7.22 REM (Marche Environmental Network). Project areas: *Montecosaro* and *Ripa Bianca* - threats and opportunities

Within the case studies coming under theme 3, we have identified two areas: Montecosaro and Ripabianca. These are situated along the branches of the "comb" that is typical of the entire regional territory, morphologically distinct from the inland hills, and marked by linear settlement development in the valley floor. The rural area is closely related to the invariants of the ecological network and to the hydrographical network. Where the orography favors it, occurrences of potential joining together of the productive/commercial fabric is seen (Fig. 8.7.22).

In conclusion, the REM project becomes the first activation of a complex and general reorganization project for the quality of the Adriatic city. The arrangement of the city is introduced starting from the environmental matrix. New **city/nature relationships** arise from this, significant not only as they increase the possibility for the displacement of plants and animals, but also because they provide spatial anchoring for actions that are more precisely directed at increasing **quality of life**. The case study in Box 8.7.1, linked to the consumption of land in the Chienti Valley and drafted in agreement with the REM project, is an eloquent example of this.

8.7.1 Containment of Land Consumption in the Lower Chienti Valley[18]

A significant case study is the Lower Chienti River Valley (Fig. 8.7.23). This is a territory with a population of 180,337 inhabitants, an average density of 452 inhabitants/km^2, in a total area of 399.35 km^2 (Fig. 8.7.24).

A knowledge framework, made of traditional thematic maps and advanced geographic information system analysis, produces a comprehensive framework for the knowledge of this area and the study of the quantitative and qualitative phenomenon of urban sprawl (Figs. 8.7.25–8.7.27).

The interpretative framework faces a range of issues related to:

- The impacts on the environment and population (Fig. 8.7.28).
- The different forms of development of the city (Fig. 8.7.29).
- The perception of the landscape in areas of urban sprawl (Fig. 8.7.30).

The strengths, weaknesses/limitations, opportunities, and threats analysis (Fig. 8.7.31) synthesizes the whole analytical and interpretative frameworks and leads to the final outcome: the intermunicipal strategic plan (Fig. 8.7.32).

[18] Andrea Renzi and Ilenia Pierantoni have further developed their thesis on "Containment of Land Consumption in the Lower Chienti Valley" (supervisors: Michele Talia, Valeria Di Palma, Scuola di Architettura e Design, 2009-2010) for their research on "urban sustainability" at the University of Camerino.

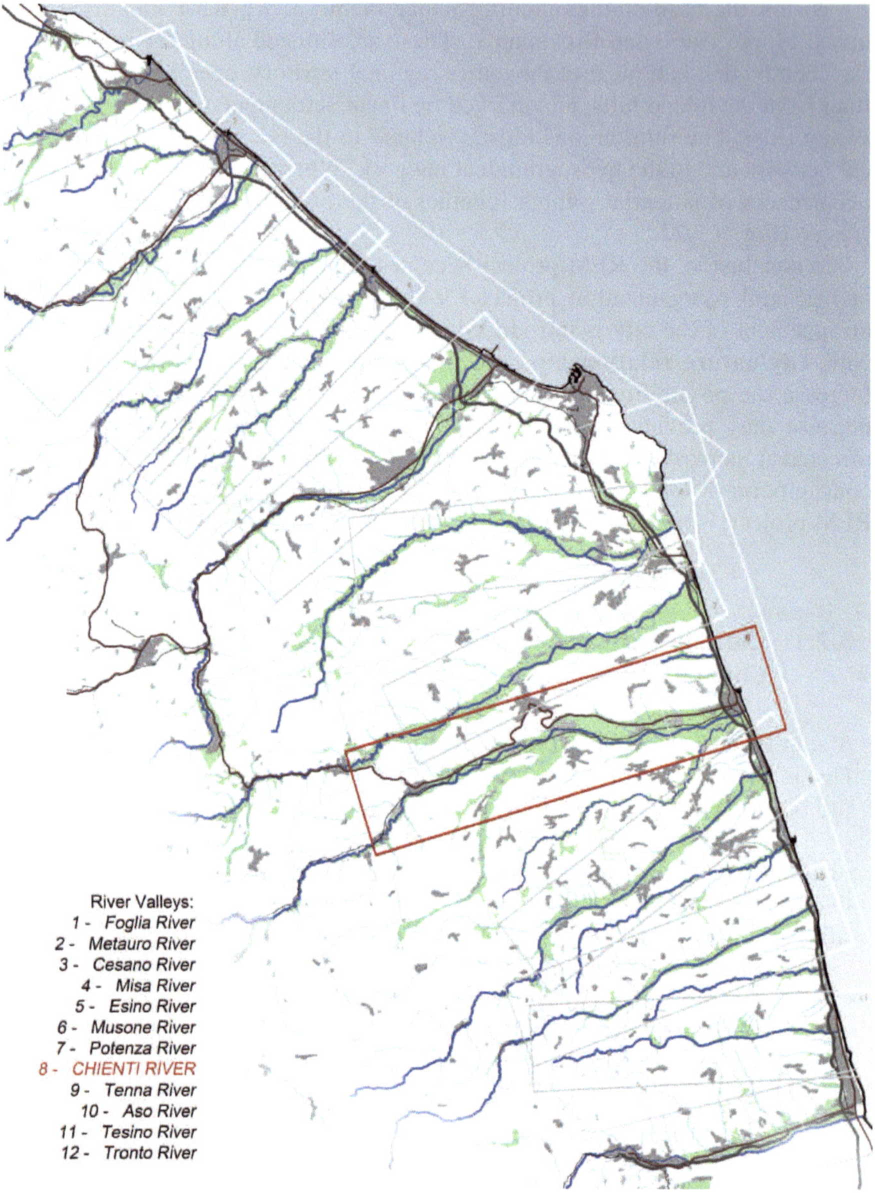

Fig. 8.7.23 The *comb system* of the Adriatic cost - Marche Region. The orthogonal system of environmental continuity of the river network to the Apennines and to the sea characterizes this region. The major components of bioconnectivity intersect or overlap with urban components in the areas in which the sprawl phenomenon is more evident. Significant is this closer examination of the case of the Chienti Valley conurbation, in which there is a deep correlation between environmental network and urban sprawl

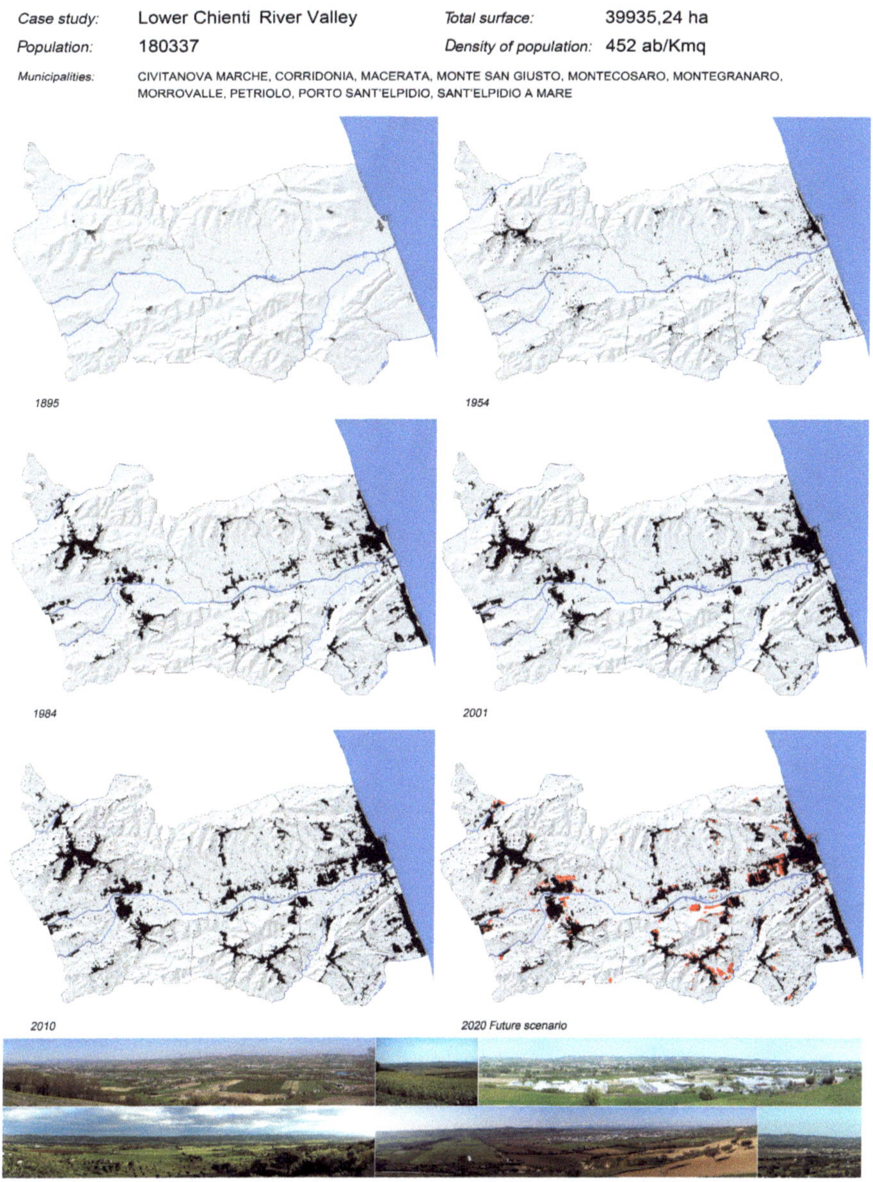

Fig. 8.7.24 The Lower Chienti River Valley. Land consumption. Analysis of urban development from 1895 to a future scenario (2020). Significant in this case is urban development from the second half of the last century up to the present, in particular in the years between 1984 and 2010

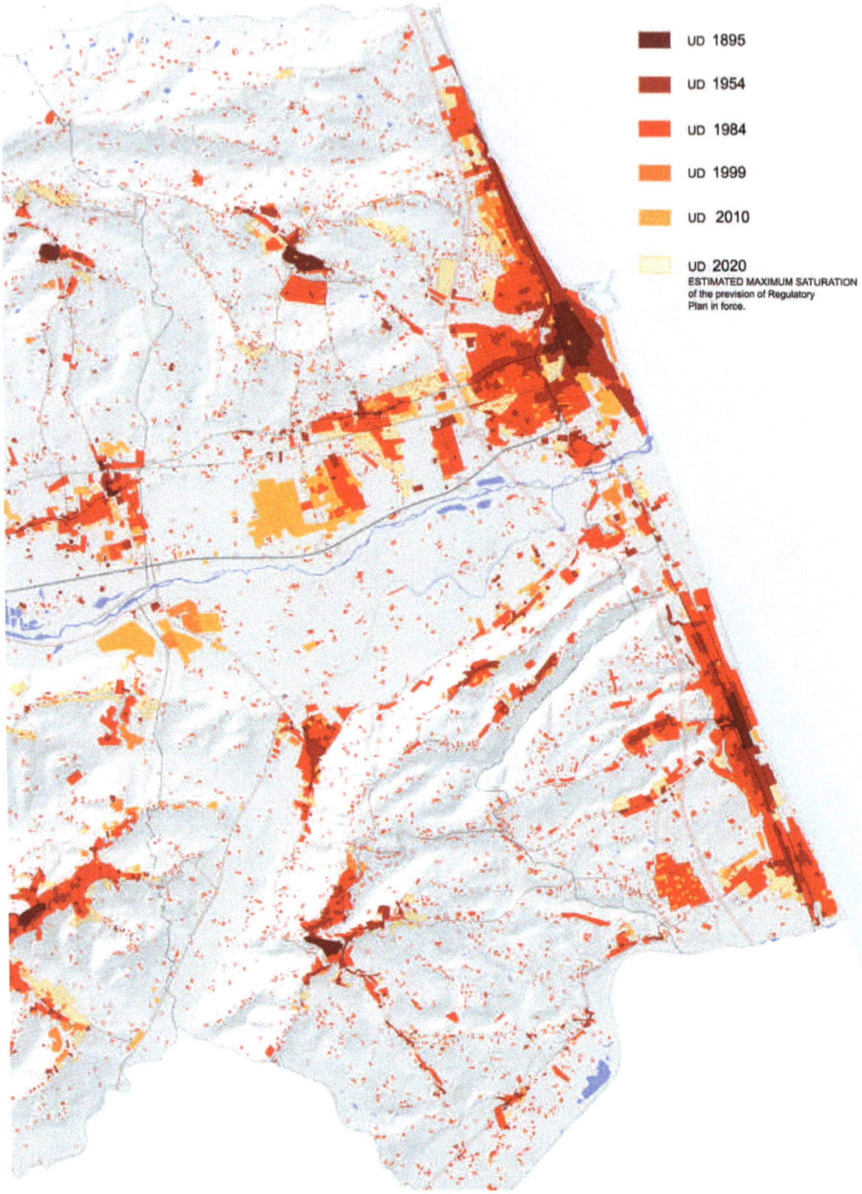

Fig. 8.7.25 Analysis of the urban development in a closer view. If we look at the size and modalities of urban growth, it is important to note that the expansions are closer and closer to the Chienti river network and to the structural elements of the ecological network

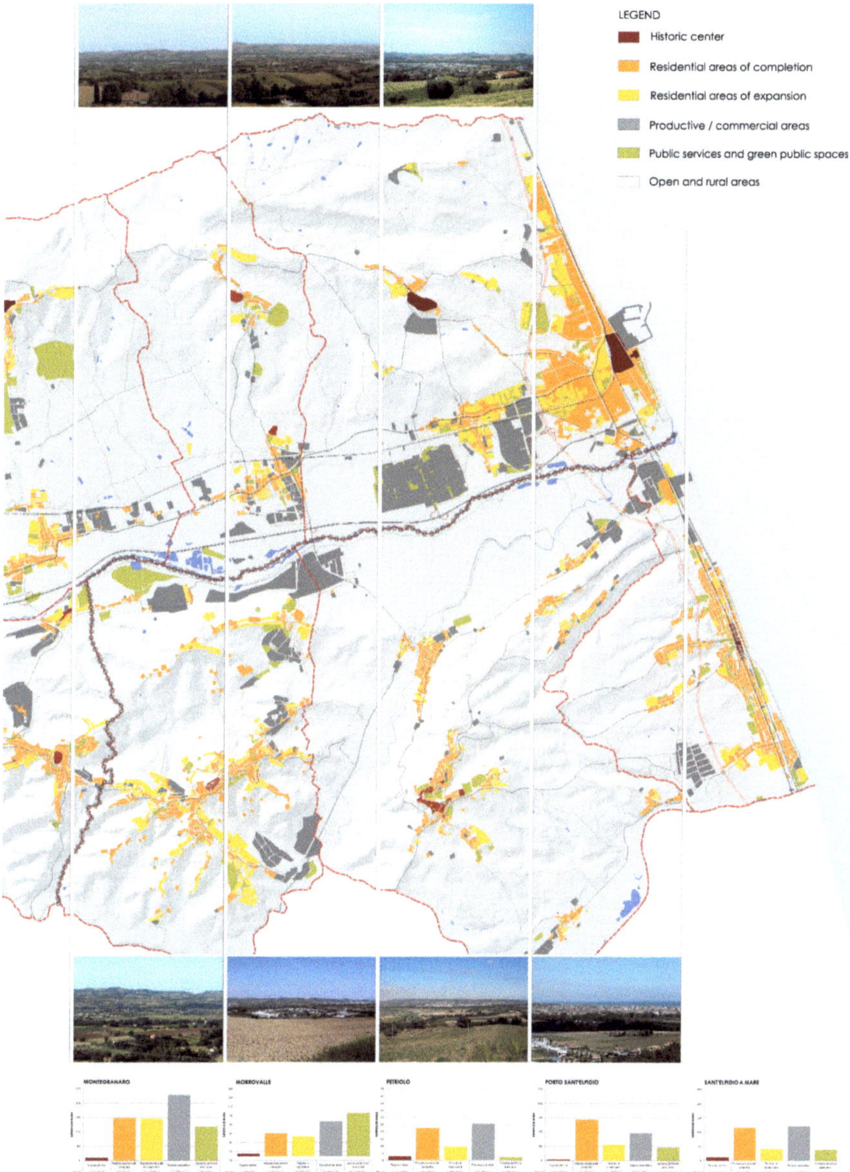

Fig. 8.7.26 The current regulatory municipal plans and the land use impacts. This analysis allows to understand how, in this new close relation between the river network and urban expansions, the presence of production and commercial areas is relevant

Fig. 8.7.27 Analysis of the expected growth resulting from the eventual implementation of the regulatory municipal plans. In the case of complete implementation of the regulatory municipal plans residential and industrial areas would be much higher than the real needs of the local communities and economies, with direct impacts on the structural elements of the environmental network, especially on the Chienti river

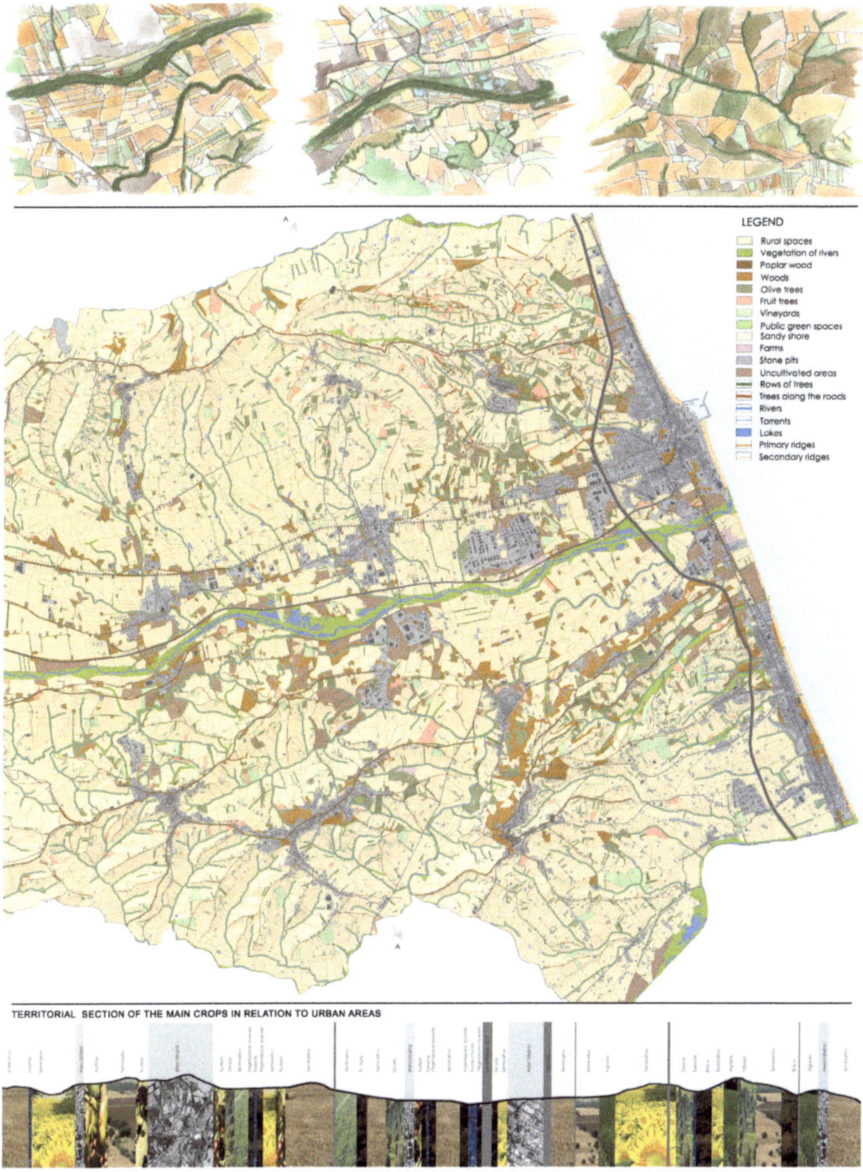

Fig. 8.7.28 Analysis of the environmental network and its relationship with urban settlements. When the environmental network, from its structural elements (vegetation formations in the river network, forest stands, and linear elements of the agrarian landscape), penetrates with thinning branches into the urban sprawl, it intersects a succession of different, repeated, and fragmented uses: agricultural, industrial, residential, infrastructure, etc.

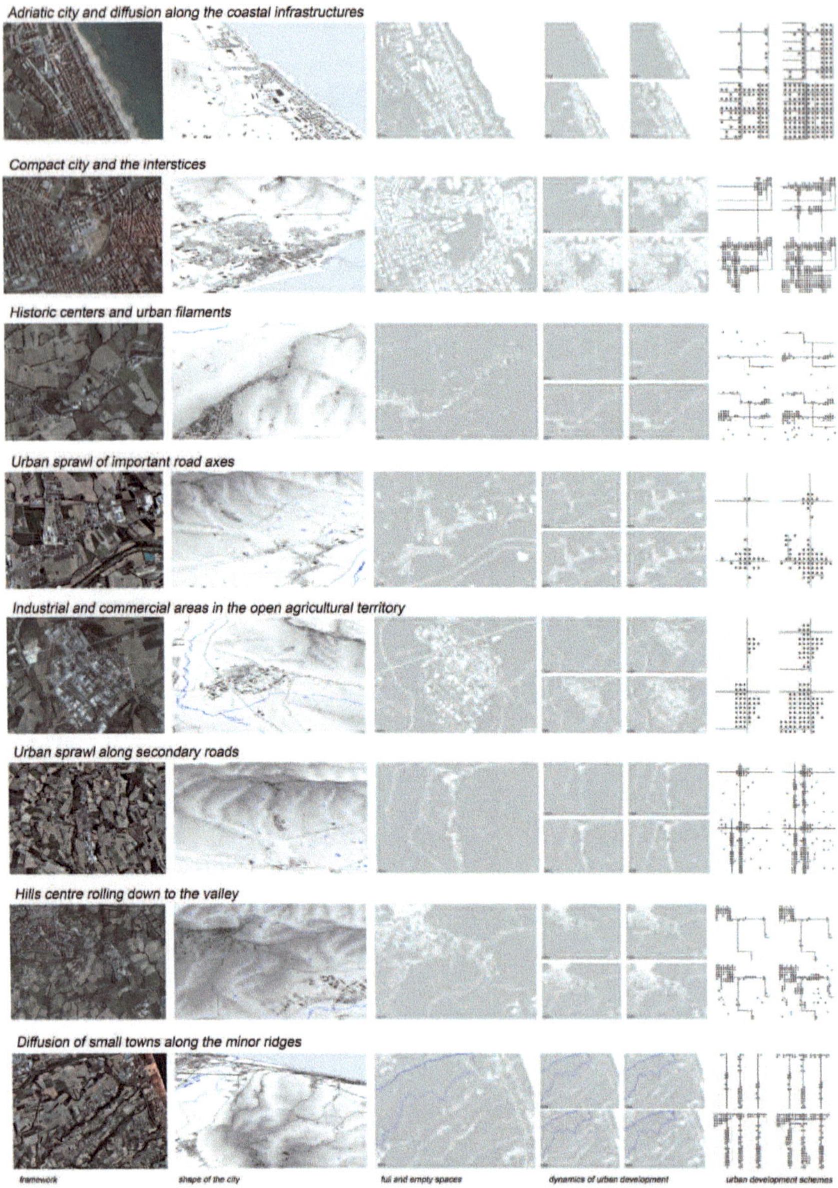

Fig. 8.7.29 The development of the contemporary city. Analysis of urban growth. The environmental network, in its several nodes and connections, intersects and penetrates the contemporary city, in its various phases of growth and consolidation: for example in the urban development along the Adriatic coast, or along important roads axes, or along ridges, or in the urban development in the open agricultural territory

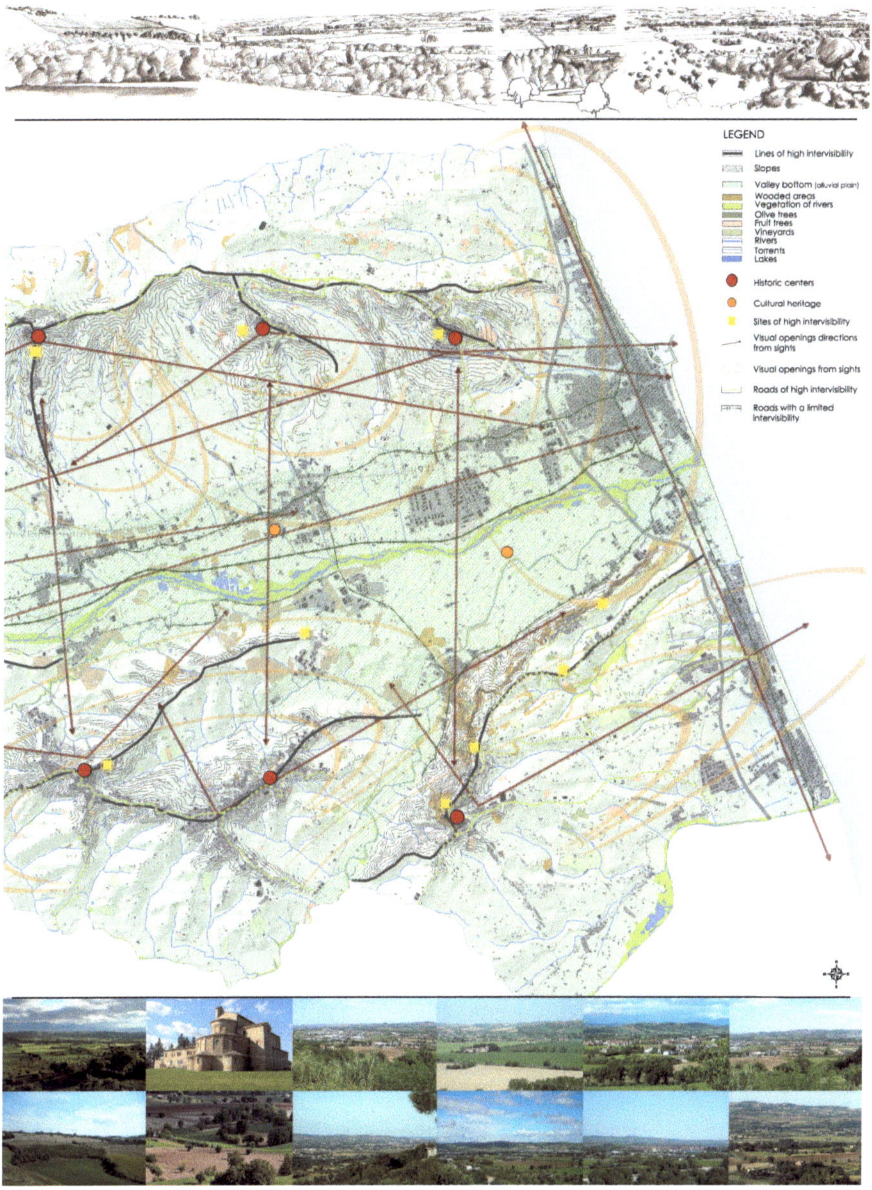

Fig. 8.7.30 Urban sprawl and the perception of landscape. In the Adriatic city, the multiple branches of environmental networks characterize the different landscape units in which the sprawl phenomenon is relevant. This green patchwork tends to become the new structure of large settlement contexts

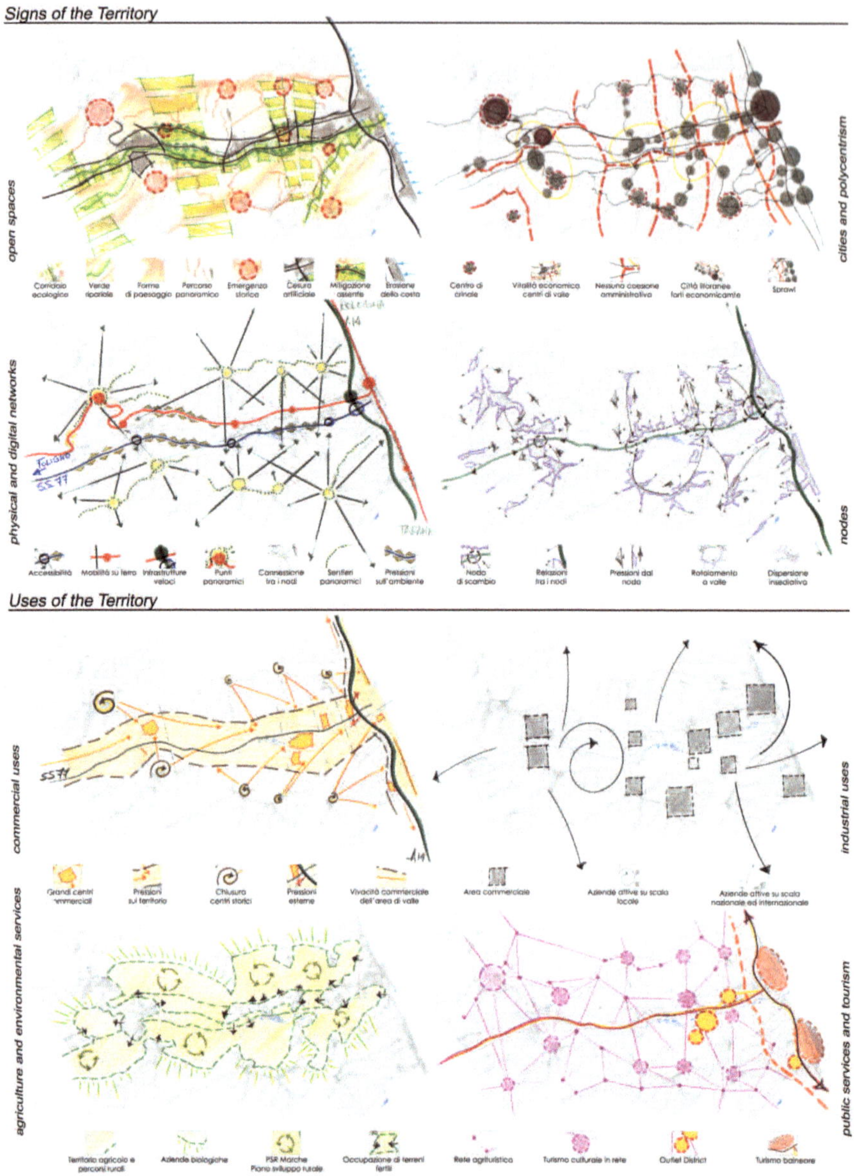

Fig. 8.7.31 SWOT analysis. This analysis allows to define Strengths, Weaknesses, Opportunities, and Threats (SWOT), with particular attention to the relationship between urban components and environmental networks, for each of the following themes: open spaces, physical and digital networks, cities and polycentrism, nodes, commercial uses, industrial uses, agriculture and environment, public services and tourism

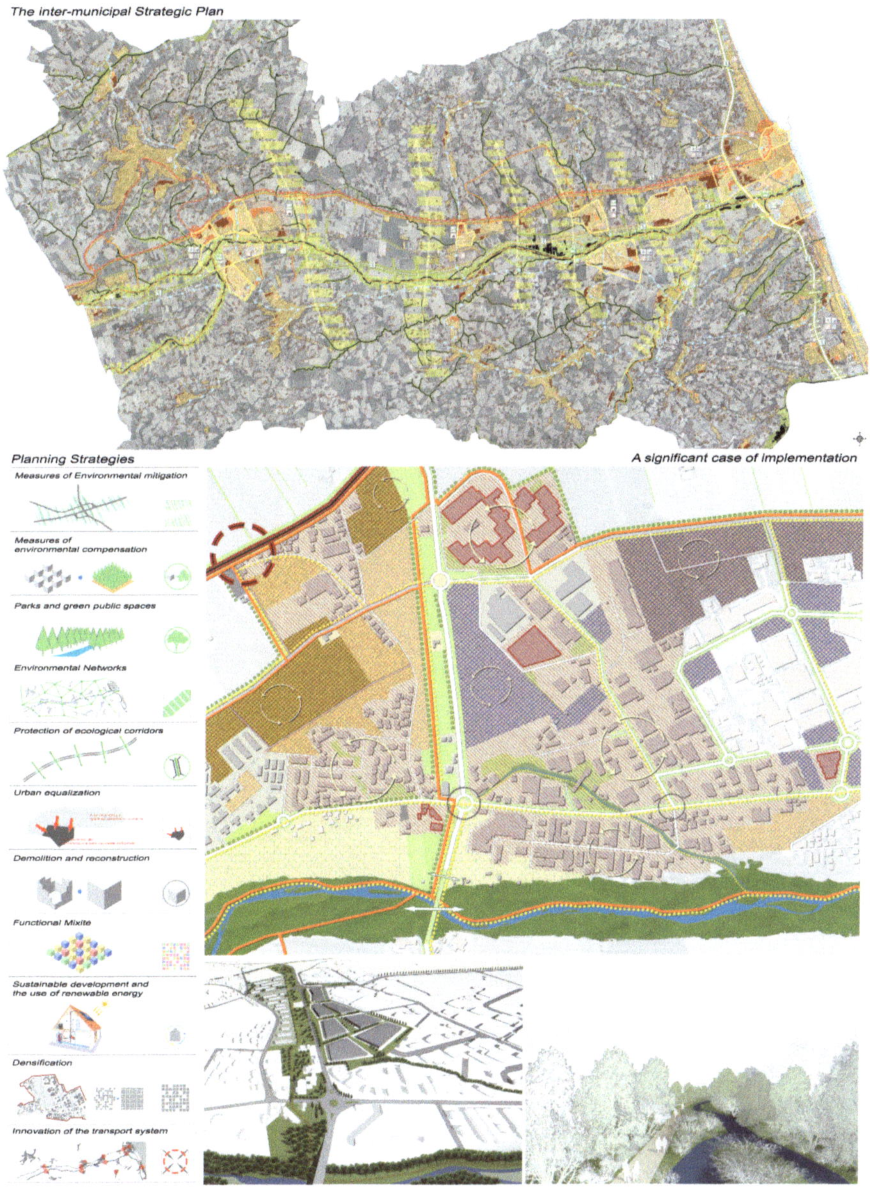

Fig. 8.7.32 The intermunicipal strategic plan. Through the environmental structure, new contacts are identified between the river network (in some parts deeply transformed by urbanization) and historical centers (from which the contemporary urban development took its origins). The implementation of this new organization and regeneration based on greenways also affects the following strategies for urban areas: measures of environmental mitigation and compensation; quantitative and qualitative improvement of parks and green public spaces; urban equalization; demolition and reconstruction processes; sustainable development and the use of renewable energies; densification; innovation of the transport system, etc.

Afterword

Giorgio Osti

This afterword is organized in two parts; one is a kind of summary concerning the topics that emerge from this book. The other is a perspective concerning the points that, in my opinion, can lead to further discussion and developments in land planning.

Regarding the summary, there are three main points:

1. The landscape serves as a tool of mediation between nature and culture, physics and *techne*.
2. The authors and editor have endeavored to bring together numerous disciplines to achieve better planning capacity.
3. The network acts as a powerful metaphor of urban development and as an organizational device for planning.

My own perspective can also be summarized in three points:

1. The distinction between the concepts of environment, network, and community is an interesting **research field** and ought to be developed further.
2. Common ground should be found in the debate on the connection between planning and **land policy implementation**, in the mismatch between decision-making and the hierarchy of objectives.
3. The knowledge of the relationship between **lifestyle** and **land use** is important, and goes beyond degrowth.

Reading this book, edited by Massimo Sargolini, is an exciting adventure across many different fields; urban planners really explore the **complexity of space** and do so for reasons beyond just a need to study it in its completeness. Urbanism is not only a pervasive morphology of space; it is not just a corresponding discipline; it is also an angle from which to look at the world. Thus, it is not unexpected that the book begins with philosophical thoughts and arrives at the end with the call for a universal approach encompassing all elements of society. Along this path, all of the conceptual tools needed to form a precise and complete picture of land planning are provided.

M. Sargolini, *Urban Landscapes*,
DOI: 10.1007/978-88-470-2880-7_9,

The many aspects touched on by Sargolini and the other authors provide us with a rich variety of food for thought. The breadth of aims contained in the book is remarkable. All of them can be linked to philosophical and ethical visions: living in a good way, creating a supportive community, respecting nature. This high quality of life—the crux of the book in my opinion—passes through a special social/environmental feature: the **landscape**. The landscape is not only, as we argue later, a cognitive device in the hands of planners, but also an ideal for the quality of life. Like the cosmos in the ancient period, the cloister in the Middle Ages, and the planned town in the Renaissance, the landscape is, according to Simmel [1], a typical product of modern life because of its mediation between subject and nature. It is the new cosmology that includes the traditional dilemma between diversity and uniformity, which has now become very urgent with the claims of ethnic migrants throughout Western countries. Furthermore, focusing on the landscape reassures us that another dilemma of the modern era can be overcome, that between nature and artifact.

How is this possible? It is a matter of coevolution: when human settlements were able to transform themselves according to local resources, an extraordinary equilibrium was established that produced a durable and agreeable environment. The former adjective indicates the ecological side of matter and the latter the aesthetic one. A beautiful environment is a condition for creating a virtuous cycle between attachment to **territory** and care for it. Such a cycle has been broken many times in many places. Recovering or maintaining a functional and pleasant environment is not easy. But this is the dramatic task we call on the planners to handle.

This introduces the second main point of the book: the **method**. In this case, Sargolini and the other authors embrace a holistic idea, or, in less philosophical terms, complexity theory. The way forward in knowledge of the landscape is the **interconnection of disciplines**: open, ongoing, dignified peer dialogue among different methods of analysis, no matter if one is more codified than another. They also stress that common knowledge and lay people must contribute to the planning process. The appeal for participation and deliberative democracy is clearly and appropriately made in the book. Of course, the path toward fair and competent participation is very long and full of obstacles, starting from the subtle hostility of institutions to grassroots movements, the reduced capacity of people to mobilize participation resources due to time and money, and, not least of all, the obscurity of evaluation procedures put in place in the experiences known in the literature. In the book, a special box is dedicated to evaluation, which is, along with transparency, the precondition for any genuine form of participation.

The third point of synthesis is the network. The title of the book is eloquent: the network perspective is a way to save both actor and system, willing people and influential structures, the subject, and common values. This is possible by focusing attention on the links between nodes—persons or organizations—rather than on the internal features of the nodes. The "reality" is, first of all, the set of relationships among actors. The form and strength of the ties can vary hugely, but the final product is the resulting bundle of relationships called a "**network**."

The application of such a device highlights the social systems and the socio-spatial systems in particular. Towns, districts, regions, parks, domains—all spatial bodies—work together in a network: the hierarchy is not marked, but is determined by the frequency of contacts rather than by the rigid, static rules of formal organization. Interdependence is the rule, thus emphasizing the prevalence of circular processes instead of linear ones. The network can be seen as a small revolution in the way organized space is represented and for the disciplines that study it.

In the research fields profiled in this book, the network appears as a suitable tool for at least two reasons: it is capable of graphically representing the **urban structure** created in the post-Fordist era (sprawl, administrative decentralization, industrial delocalization, multiple destination mobility, and so on); and it also represents very well the idea of **teamwork**, reflecting the interdisciplinary approach. The real world and students of it have a sort of structural homology, i.e., the network is the common framework. Furthermore, it should be mentioned that the ideal network could be a tool for ensuring a certain degree of nature protection in highly urbanized environments, which is the case for the majority of the Italian territory.

Because post-Fordist urbanization has fragmented the ecosystems, and we know that wild species need a large area to thrive, the remedy is to organize conservation through networks or corridors capable of progressively connecting together smaller seminatural areas. It is impossible to design or forecast the recentralization of human settlements and services to achieve more extensive wild areas. Faced with this limit, the idea of connecting residual natural areas together with ecological corridors perfectly fits within the network approach. Some ecological services could also be organized as a network, for example, freshwater could be provided not only through a centralized source, but also drawn from many interconnected wells and springs. Electricity is not really an ecological service, but smart grids are inspired by the powerful idea of the network, either as sources or for provision.

Network is such a meaningful and multifunctional concept that it risks becoming a sort of tool for all trades. This introduces the critical points highlighted in the book. The relationships and analytical differences between communities and networks also ought to be developed more at the ecological and sociological level. In both cases, community has a basic area element. "Area" indicates a surface, a geographical unit, and by extension, a specific approach to land planning. On the other hand, the network is made of points and links; what is in the middle of the web is not important or left blank.

Community is, generally speaking, a residential unit, i.e., an area where populations live constantly. Of course, the mobility of objects and people has modified the closeness of community, but not to the point that it nullifies the value and effects of living nearby, as the deterritorialization hypothesis leads us to believe. In the book, this idea of virtual communities substituting physical communities is mentioned. In other words, the world is made of networks; people live in contact with many other people without regard for their location in space. There are two points of reflection regarding this position that can lead to further discussion and developments.

The first concerns the real world. Even if mobility and communication are very diffuse and significant, the space of the place, to use Castell's terminology [2], maintains its priority in the sense of life as well as in daily life. In other words, people need and practice community. It can be very weak, poorly organized, even unhealthy, but some level of community life is a basic need for every human being. Nobody can live only within the space of flux. This is even truer for nonhuman species, whose grounding in one stable place is vital.

The second point to investigate concerns the **normative dimension**. All researchers stress ancient roots, territorial belonging, and attachment to a residential community as powerful tools in achieving a better quality of life. These tools are also useful for environmental protection, since people who identify with their communities are attached to the landscape and the environment. This normative idea often arises in the book, thus raising some questions: Is it possible to attach to a network? Is the landscape based on the landscape? How can we connect the community to the network? It is important that the complexity of the problem does not lead to confusion between networks and communities, since this could lead to mistakes in the representation of space and in land planning. The former are more virtual in nature, i.e., they can live on very thin ground; the latter have a deeper connection with the land.

A further difference between community and network concerns **values**. A community usually implies shared values. In this sense, it exercises the pressure to conform on its members, as Portes [3] said when describing ethnic communities. A network implies only the common rules of **communication** and **exchange**. Because of this, it is much preferred in the modern era. Uniformity is not required, and radically different people are allowed to dialogue. This represents the urban condition well, where different ethos can live side-by-side because their points of reference are elsewhere. In turn, it explains the emphasis on borders in community studies and the methodological difficulty in discovering where the edges of a network lie. The ideal community is a clique whose points have no link with the external world. The ideal network, on the contrary, is a continuous solution. The chain of relationships is basically endless.

For **land planning**, referring to communities or networks is equally easy. For example, both transport systems and commercial areas must be described in terms of fluxes and sinks. You move from one place to another or you buy in a central location because you know it and have functional links with the area surrounding the point you are going to. The concepts of "activation space" or "territorial competence" in geographical mobility indicate an actor mixing an area with a web perspective. The case of landscape is a bit different. Such representation is basically based on area. Tracks, paths, and lines can be distinguished in the landscape but as parts that are organically linked to a comprehensive entity. Otherwise, lines come to represent the borders of a specific landscape. For this reason, landscape belongs to the language of community rather than network.

Similar observations can be made for the environment. Ecosystems have a substantial continuity in space; they are more similar to areas than to networks. Network abstraction appears to be an expedient of planners for finding a remedy to habitat fragmentation. In no way can networks be presented as a model of nature conservation. Large, contiguous reserves are the best way to maintain biodiversity. Also, the logic of buffer zones around protected areas is a clear example of distance from a network pattern. Of course, in deeply urbanized territory, ecological networks are the only viable subordinate model.

One more issue concerns the need to develop more how the movement from land planning to the **implementation of policies** happens. To put the question in another way: the planning philosophy is good (multidisciplinary, bottom-up, systemic, and so on) and the products, the plans, are good as well. There is awareness of how fragile the relationship between planners and politicians is, and of how important the effective implementation of land policies is. A set of **indicators** provides the right example. Many lists of indicators for assessing the state of the world are mentioned; a depth study will focus on the compatibility issues among environmental indicators developed at different measure scales. Formal adaptation among them is possible by providing synthetic indices. Analyzing the concrete use of such indicators is necessary; this is a political process, requiring the tools of policy analysis.

The problem can be formulated in a more radical way, as appears in this book: land planning needs a **self-monitoring activity**, a capacity for the planning system to observe itself. Its importance arises from the fact that this activity is mentioned at different points in the book, even though a policy analysis specialist has not been involved. Even if every function entails a specialist, the problem is not limited to including a further expert in the author pool. Monitoring means that inside the core discipline of landscape ecology a set of tools and procedures must be developed to include political skills. And this is precisely the point: such skills have to come from inside the planning system; in other words, they must be self-monitoring skills.

The lack of a strong relationship with the implementation phase also comes from another shortcoming of our disciplines when facing systemic objects, such as landscapes, environments, and societies. It is the lack of hierarchy, i.e., determined priorities. The method affects the results; an emphasis on interdependence, which is typical of networks, reduces the capacity to determine the most urgent points. One compromise is to use the metaphor of a chain, in which there is interdependency but also linearity, thus limiting the number of intervening factors.

The last issue concerns the general debate on **sustainability**. Of course, sustainability is a necessary requirement for every intervention in the territory. The book starts with a quotation by Latouche, the famous writer of degrowth. Such a radical vision of the world is now being challenged for two reasons. One is the insufficient elaboration in the **degrowth doctrine** regarding redistributional imperatives, i.e., which social classes have to downshift. The second source of criticism concerns the impasse in facing the real degrowth that the GDP has registered in these years of economic crisis. Is this the kind of change that Latouche and his supporters want? Evidently, the distinction between economic dematerialization and GDP decline, and the distinction between relative and absolute downshift have not been clarified sufficiently. Nevertheless, the gaps in the degrowth doctrine cannot be an excuse for avoiding the basic problem of future scarcity. The book is rich in analysis on sustainability in many fields. Some are more urgent than others, and this recalls the need to find a hierarchy among environmental objectives.

But the real issue is the absolute voluntary **reduction of the use of crucial resources** such as water, energy, and land. They are relatively abundant at the moment and with an "instant" (present-driven) attitude, the perception of scarcity is difficult. How were the disciplines in this book able to communicate the minimal sense of future helpful for understanding the limits of resources? A realistic account of the number of years before oil and freshwater are exhausted is difficult at the scientific level, as demonstrated by the Club of Rome prediction. Evaluations are linked to cultural assumptions on risks, human nature, and other people's degree of liability.

Yet again this calls for a new integration between the physical sciences and cultural disciplines. Some decades ago, Mary Douglas showed that **risk perception** is linked to the structure and **cohesion of society**. It is probable that we are currently facing a new age of the **perception of scarcity**. Otherwise, how can we explain why China, unlike European countries, is buying so much land in Africa? **Land grabbing**, as this phenomenon is called, raises a lot of criticism. Nevertheless, this hoarding, which often does not lead to rational exploitation of the land, responds to a specific culture of future, a mix of science, emotions, and memory of past scarcities.

The book correctly tries to maintain such an interdisciplinary, cross-society mix. The problem is that many people do not accept or do not understand such an all-encompassing view. Landscape analysis is likely the emblem of an enduring difficult dialogue between land planning specialists and the broader community. The book admirably presents both sides of the conflict, as Sargolini comments on landscape in these terms:

> *Where the landscape is poorly integrated with the living community, processes should be triggered to render collective a sense that otherwise would remain completely subjective and generate very different and contrasting results.*

However, the task is at least reciprocal: reducing the disintegration of community, an enduring problem since the times of the Chicago Ecological School, and reducing the fragmentation and aphasia of land experts, an endemic risk, even in the most motivated groups.

References

1. Simmel G (1913) Philosophie der Landschaft, Gesamtausgabe. Vol. 12; trad. it. Saggi sul paesaggio, Roma, Armando, 2006
2. Castells M (1996) The rise of the network society. The information age. Oxford, Blackwell; trad. it. La nascita della società in rete. Milano, Università Bocconi, 2002
3. Portes A, Sensebrenner J (1993) Embeddedness and immigration: Notes on the social determinants of economic action. American Journal of Sociology 98:1320-1350

MIX
Papier aus verantwortungsvollen Quellen
Paper from responsible sources
FSC® C105338

If you have any concerns about our products,
you can contact us on
ProductSafety@springernature.com

In case Publisher is established outside the EU,
the EU authorized representative is:
Springer Nature Customer Service Center GmbH
Europaplatz 3, 69115 Heidelberg, Germany

Printed by Libri Plureos GmbH
in Hamburg, Germany